Lectures in Mathematics
ETH Zürich
Department of Mathematics
Research Institute of Mathematics

Heinz-Otto Kreiss
Hedwig Ulmer Busenhart

Time-dependent Partial Differential Equations and Their Numerical Solution

Springer Basel AG

Authors' addresses:

Heinz-Otto Kreiss
Department of Mathematics
University of California Los Angeles
Los Angeles, CA 90095-1555
USA

Hedwig Ulmer Busenhart
Scheuchzerstr. 18
8006 Zürich
Switzerland

2000 Mathematical Subject Classification 65MXX

A CIP catalogue record for this book is available from the
Library of Congress, Washington D.C., USA

Deutsche Bibliothek Cataloging-in-Publication Data
Kreiss, Heinz-Otto:
Time-dependent partial differential equations and their numerical solution /
Heinz-Otto Kreiss ; Hedwig Ulmer Busenhart. - Basel ; Boston ; Berlin :
Birkhäuser, 2001
(Lectures in mathematics : ETH Zürich)
ISBN 978-3-7643-6125-9 ISBN 978-3-0348-8229-3 (eBook)
DOI 10.1007/978-3-0348-8229-3

ISBN 978-3-7643-6125-9

Originally published by Birkhäuser Verlag in 2001

Printed on acid-free paper produced from chlorine-free pulp. TCF ∞

ISBN 978-3-7643-6125-9

9 8 7 6 5 4 3 2 1

Contents

Preface

These notes are based on the following two books:

1. Heinz-Otto Kreiss and Jens Lorenz: "Initial-Boundary Value Problems and the Navier-Stokes Equations", *Academic Press, Inc.*, 1989
2. Bertil Gustafsson, Heinz-Otto Kreiss and Joseph Oliger: "Time Dependent Problems and Difference Methods", *John Wiley & Sons, Inc.*, 1995.

They are concerned with the theory of time dependent partial differential equations and their numerical solution by difference approximations. Our intent has been to highlight the main ideas. To faciliate a more detailed study we will make more specific references at the end of each chapter.

We will refer to the literature only sparingly since an extensive list of references can be found in the above books.

Chapter 1

Cauchy Problems

1.1 Introductory Examples

Example 1. We consider the initial value problem

$$\begin{aligned} u_t(x,t) + u_x(x,t) &= 0, \qquad x \in \mathbb{R}, \quad t \geq 0, \\ u(x,0) &= f(x), \quad x \in \mathbb{R}, \end{aligned} \tag{1.1.1}$$

where $f(x) = f(x + 2\pi)$ is a smooth 2π-periodic function. To begin, we assume that

$$f(x) = e^{i\omega x} \hat{f}(\omega)$$

consists of only one wave, where $\hat{f}(\omega)$ denotes the Fourier transform of $f(x)$. In order to construct a solution of the same type we choose the ansatz

$$u(x,t) = e^{i\omega x} \hat{u}(\omega, t). \tag{1.1.2}$$

Substituting Equation (1.1.2) into Equation (1.1.1), we obtain the ordinary differential equation

$$\begin{aligned} &\hat{u}(\omega,t)_t + i\omega \hat{u}(\omega,t) = 0, \\ &\hat{u}(\omega,0) = \hat{f}(\omega), \end{aligned}$$

which is called the Fourier transform of Equation (1.1.1) and has the solution

$$\hat{u}(\omega,t) = e^{-i\omega t} \hat{u}(\omega,0).$$

It follows that

$$u(x,t) = e^{i\omega(x-t)} \hat{u}(\omega,0) = f(x-t)$$

is a solution of (1.1.1). Now consider the general case

$$f(x) = \frac{1}{\sqrt{2\pi}} \sum_{\omega=-\infty}^{\infty} e^{i\omega x} \hat{f}(\omega).$$

By the superposition principle

$$u(x,t) = \frac{1}{\sqrt{2\pi}} \sum_{\omega=-\infty}^{\infty} e^{i\omega(x-t)} \hat{f}(\omega) = f(x-t)$$

is a solution of (1.1.1). We denote by $\bar{f}$ the conjugate complex value of f. We define the L_2 scalar product and norm by

$$(f,g) = \int_0^{2\pi} \bar{f} g dx, \tag{1.1.3}$$

$$\|f\| = (f,f)^{1/2} \tag{1.1.4}$$

Using Parseval's relation

$$\sum_{\omega=-\infty}^{\infty} \left|\hat{f}(\omega)\right|^2 = \|f(\cdot)\|^2, \tag{1.1.5}$$

we obtain for every fixed t

$$\|u(\cdot,t)\|^2 = \sum_{\omega=-\infty}^{\infty} \left|e^{i\omega t}\hat{f}(\omega)\right|^2 = \sum_{\omega=-\infty}^{\infty} \left|\hat{f}(\omega)\right|^2 = \|f(\cdot)\|^2.$$

The amplitude of the solution of the original problem does not increase with time. The problem (1.1.1) is said to be *norm conserving.*

In the following examples we restrict our attention to simple wave solutions.

Example 2. Next consider the heat equation

$$\begin{aligned} u_t(x,t) &= u_{xx}(x,t), \quad x \in \mathbb{R}, \quad t \geq 0, \\ u(x,0) &= f(x), \qquad x \in \mathbb{R}. \end{aligned} \tag{1.1.6}$$

The same ansatz as in Example 1 leads to the ODE

$$\hat{u}_t(\omega,t) = -\omega^2 \hat{u}(\omega,t),$$

which is solved by

$$\hat{u}(\omega,t) = e^{-\omega^2 t} \hat{u}(\omega,0).$$

Hence the solution of the heat equation is

$$u(x,t) = e^{i\omega x - \omega^2 t}\hat{u}(\omega,0).$$

For large $|\omega|$, the solution decays rapidly.

Example 3. The linearized flame front equation is given by

$$u_t(x,t) + u_x(x,t) + u_{xx}(x,t) = -\nu u_{xxxx}(x,t). \tag{1.1.7}$$

Again by the ansatz (1.1.2) we obtain

$$\hat{u}_t(\omega,t) = \left(-i\omega + \omega^2 - \nu\omega^4\right)\hat{u}(\omega,0)$$

with the solution

$$\hat{u}(\omega,t) = e^{\left(-i\omega+\omega^2-\nu\omega^4\right)t}\hat{u}(\omega,0).$$

For $|\omega|^2 > \nu^{-1}$, $\nu\omega^4$ dominates and the amplitude of $\hat{u}(\omega,t)$ decays. For $|\omega|^2 < \nu^{-1}$, the amplitude grows. The growth rate of the amplitude is maximal at $\omega^2 = \frac{1}{2}\nu^{-1}$ but still independent of ω and

$$|\hat{u}(\omega,t)| = e^{\frac{1}{4\nu}t}|\hat{u}(\omega,0)|, \quad \omega = \sqrt{1/2\nu}.$$

Example 4. The backward heat equation

$$u_t(x,t) = -u_{xx}(x,t) \tag{1.1.8}$$

behaves even worse. The amplitude of the solution is

$$\hat{u}(\omega,t) = e^{\omega^2 t}\hat{u}(\omega,0).$$

There is no bound for the growth. The problem is not well posed. The idea to restrict the numerical calculation to just a few low frequencies does not work out, because the rounding errors will always lead to high frequency components.

1.2 Well-Posedness

In this section we want to give an operational definition of well-posedness for the Cauchy problem

$$\begin{aligned} u_t(x,t) &= P(\partial/\partial x)\,u(x,t), & x &\in \mathbb{R}^d, \quad t \geq 0, \\ u(x,0) &= f(x), & x &\in \mathbb{R}^d. \end{aligned} \tag{1.2.9}$$

We start with some notational remarks. $P(\partial/\partial x)$ is a differential operator of order m

$$P(\partial/\partial x) = \sum_{|\nu|\leq m} A_\nu D^\nu,$$

with constant complex coefficient matrices A_ν and ν is a multi-index, i.e. a vector with non negative integers as components. Its order is $|\nu| = \nu_1 + \cdots + \nu_d$. Thus D^ν is defined by

$$D^\nu = \frac{\partial^{|\nu|}}{\partial x_1^{\nu_1} \cdots \partial x_d^{\nu_d}}.$$

We assume that the initial function $f(x)$ is smooth and 2π-periodic in all space directions and thus can be expanded into a rapidly converging Fourier series

$$f(x) = \sum_{\omega=-\infty}^{+\infty} e^{i\langle\omega,x\rangle} \hat{f}(\omega),$$

where $\langle\omega, x\rangle = \sum_{i=1}^d \omega_i x_i$ stands for the scalar product of ω and x and ω is a vector with integer components. We observe that the application of the differential operator $P(\partial/\partial x)$ to $e^{i\langle\omega,x\rangle}\hat{f}(\omega)$ results in multiplication by the matrix

$$P(i\omega) := \sum_{|\nu|\leq m} A_\nu (i\omega_1)^{\nu_1} \cdots (i\omega_d)^{\nu_d}.$$

The matrix $P(i\omega)$ is called the *symbol* of $P(\partial/\partial x)$. We now give a definition of well-posedness in terms of the symbol $P(i\omega)$.

Definition 1.2.1 *The Cauchy problem (1.2.9) is well posed if there are constants K, α independent of ω such that*

$$\left|e^{P(i\omega)t}\right| \leq Ke^{\alpha t}, \tag{1.2.10}$$

for all $t \geq 0$ and all ω.

Assuming that the system (1.2.9) is well posed we can solve it by using the ansatz

$$u(x,t) = \sum_{\omega=-\infty}^{+\infty} e^{i\langle\omega,x\rangle} \hat{u}(\omega,t).$$

leading to the initial value problem for the ODE

$$\begin{aligned} \hat{u}_t(\omega,t) &= P(i\omega)\hat{u}(\omega,x), \\ \hat{u}_t(\omega,0) &= \hat{f}(\omega). \end{aligned} \tag{1.2.11}$$

The solution of equation (1.2.11) is given by

$$\hat{u}(\omega, t) = e^{P(i\omega)t} \hat{f}(\omega).$$

Thus the solution of (1.2.9) reads

$$u(x,t) = \sum_{\omega=-\infty}^{+\infty} e^{i\langle \omega, x\rangle} e^{P(i\omega)t} \hat{f}(\omega).$$

The sum converges because $\left|e^{P(i\omega)t}\right|$ is bounded due to the well-posedness of the problem. We can now prove the following theorem:

Theorem 1.2.1 *The Cauchy problem (1.2.9) is well posed if and only if*

$$\|u(\cdot,t)\|^2 \le K^2 e^{2\alpha t} \|u(\cdot,0)\|^2.$$

Proof. Using Parseval's relation we obtain the estimates

$$\begin{aligned} \|u(\cdot,t)\|^2 &= \sum_{\omega=-\infty}^{+\infty} |\hat{u}(\omega,t)|^2 = \sum_{\omega=-\infty}^{+\infty} \left|e^{P(i\omega)t}\hat{f}(\omega)\right|^2 \\ &\le \sup_{\omega} \left|e^{P(i\omega)t}\right|^2 \sum_{\omega=-\infty}^{+\infty} \left|\hat{f}(\omega)\right|^2 \\ &\le K e^{2\alpha t} \|u(\cdot,0)\|^2. \end{aligned}$$

The other direction is trivial. □

1.3 Hyperbolic Systems with Constant Coefficients

1.3.1 In One Space Dimension

Consider a first order system

$$\begin{aligned} u_t(x,t) &= A u_x(x,t), \\ u(x,0) &= f(x), \end{aligned} \qquad (1.3.1)$$

where $u = (u_1, u_2, \ldots, u_n)^T$ and A is a complex $(n \times n)$-matrix.

Theorem 1.3.1 *The Cauchy problem (1.3.1) is well posed if and only if the eigenvalues of A are real and A has a complete set of eigenvectors.*

Proof. "$\Leftarrow$" If the eigenvalues of A are real and there is a complete set of eigenvectors we can find a nonsingular transformation T that transforms A to diagonal form

$$TAT^{-1} = \begin{pmatrix} \lambda_1 & & 0 \\ & \ddots & \\ 0 & & \lambda_r \end{pmatrix} = \Lambda.$$

Therefore

$$\left|e^{i\omega At}\right| = \left|T^{-1}Te^{i\omega At}T^{-1}T\right| \leq |T^{-1}| \left|e^{i\omega \Lambda t}\right| |T| = |T^{-1}||T|,$$

since the eigenvalues of A are real. Therefore the problem is well posed with $K = |T^{-1}||T|$.

"$\Rightarrow$" We transform A to Jordan's normalform

$$TAT^{-1} = \begin{pmatrix} J_1 & & 0 \\ & \ddots & \\ 0 & & J_r \end{pmatrix},$$

where

$$J_j = \lambda_j I + D_j \quad \text{and} \quad D_j = \begin{pmatrix} 0 & 1 & & 0 \\ & \ddots & \ddots & \\ & & \ddots & 1 \\ 0 & & & 0 \end{pmatrix}.$$

Since

$$\begin{aligned} \left|e^{i\omega At}\right| &= \left|T^{-1}Te^{i\omega At}T^{-1}T\right| \\ &\geq \frac{1}{|T^{-1}|\,|T|} \max_j \left|e^{i\omega J_j t}\right| \\ &= \frac{1}{|T^{-1}||T|} \max_j \left|e^{i\omega \lambda_j t}\right| \left|e^{i\omega D_j t}\right|. \end{aligned}$$

We denote the Jordan block J_k which maximizes the term $\left|e^{i\omega J_j t}\right|$ by J and the corresponding eigenvalue by $\lambda = a + ib$. Then we can rewrite the above inequality as follows

$$\left|e^{i\omega At}\right| \geq \frac{1}{|T^{-1}|\,|T|} \left|e^{i\omega at}\right| \left|e^{-\omega bt}\right| \left|e^{i\omega Dt}\right|.$$

For $b \neq 0$, we can always choose ω such that $|e^{-\omega bt}|$ grows faster than $e^{\alpha t}$ for any α. Therefore a necessary condition for well-posedness is that the eigenvalues are real, i.e. $b = 0$. If D is a $p \times p$ block, we have

$$e^{i\omega Dt} = \sum_{j=0}^{p-1} \frac{\omega^j D^j t^j}{j!},$$

because D is nilpotent with $D^p = 0$. Thus $|e^{i\omega Dt}|$ grows like $|\omega t|^{p-1}$ and (1.2.10) can only hold if $p = 1$. Therefore all the matrices D_j must have the dimension one, which is equivalent with A having a complete set of eigenvectors. □

1.3.2 Symmetrizer

Again consider

$$\hat{u}_t(\omega, t) = i\omega A\hat{u}(\omega, t). \tag{1.3.2}$$

Assume that the matrix A is hermitian, i.e. $A = A^* = \bar{A}^T$, and look at the time development of the energy

$$\begin{aligned} \frac{\partial}{\partial t}\langle \hat{u}, \hat{u}\rangle &= \langle \hat{u}, \hat{u}_t\rangle + \langle \hat{u}_t, \hat{u}\rangle \\ &= \langle \hat{u}, i\omega A\hat{u}\rangle + \langle i\omega A\hat{u}, \hat{u}\rangle \\ &= \langle -i\omega A\hat{u}, \hat{u}\rangle + \langle i\omega A\hat{u}, \hat{u}\rangle \\ &= 0, \end{aligned}$$

i. e.

$$|\hat{u}(\omega, t)|^2 = |\hat{u}(\omega, 0)|^2, \quad \text{for all } t \geq 0.$$

Thus (1.2.10) is satisfied with $K = 1$, $\alpha = 0$. The following lemma shows that for well-posed problems one can always change the norm in such a way that the Matrix A is hermitian in the new norm defined by $|\hat{u}|_H^2 = \langle \hat{u}, H\hat{u}\rangle$.

Lemma 1.3.1 *Let A be a $(n \times n)$-matrix. There is a positive definite Hermitian matrix H such that*

$$HA + A^*H = 0 \tag{1.3.3}$$

if and only if all eigenvalues of A are purely imaginary and A has a complete set of eigenvectors. H is called a symmetrizer of A.

Proof. "$\Leftarrow$" First suppose that all eigenvalues of A are purely imaginary and that A has a complete set of eigenvectors. Then we can find a matrix T such that

$$TAT^{-1} = \begin{pmatrix} \lambda_1 & & 0 \\ & \ddots & \\ 0 & & \lambda_n \end{pmatrix} = \Lambda$$

We claim that $H = T^*T$ is a symmetrizer and check that (1.3.3) holds

$$T^*TA + A^*T^*T = T^*\left(TAT^{-1} + (T^*)^{-1}A^*T^*\right)T = T^*(\Lambda + \Lambda^*)T = 0.$$

Thus $H = T^*T$ has the desired property.

"⇒" To prove the other direction we assume that there exists a positive definite Hermitian matrix $H = S^*S$ such that

$$S^*SA + A^*S^*S = 0.$$

But then

$$SAS^{-1} + (S^*)^{-1} A^*S^* = 0.$$

Since SAS^{-1} is anti-Hermitian we can find a unitary transformation U such that

$$USAS^{-1}U^* = \begin{pmatrix} \lambda_1 & & 0 \\ & \ddots & \\ 0 & & \lambda_n \end{pmatrix}, \quad \lambda_j \text{ purely imaginary,}$$

which completes the proof. □

We now consider equation 1.3.2. If the problem is well posed, the eigenvalues of the matrix $\tilde{A} = i\omega A$ are purely imaginary and $\tilde{A}$ has a complete set of eigenvectors. Due to the lemma we can construct a matrix H such that $H\tilde{A} + \tilde{A}^*H = 0$. In the new norm $\langle \hat{u}, H\hat{u}\rangle$ we get a contraction since:

$$\begin{aligned} \frac{\partial}{\partial t}\langle \hat{u}, H\hat{u}\rangle &= \langle \hat{u}_t, H\hat{u}\rangle + \langle \hat{u}, H\hat{u}_t\rangle \\ &= \langle i\omega A\hat{u}, H\hat{u}\rangle + \langle \hat{u}, Hi\omega A\hat{u}\rangle \\ &= \langle \hat{u}, \underbrace{\left(\tilde{A}^*H + H\tilde{A}\right)}_{=0} \hat{u}\rangle = 0. \end{aligned}$$

1.3.3 Multiple Space Dimensions

The Cauchy problem for first order systems in d space dimensions has the form

$$\begin{aligned} \frac{\partial u}{\partial t}(x,t) &= \sum_{\nu=1}^{d} A_\nu \frac{\partial u}{\partial x_\nu}(x,t), \quad x \in \mathbb{R}^d, \quad t \geq 0, \\ u(x,0) &= f(x), \qquad\qquad x \in \mathbb{R}^d. \end{aligned} \tag{1.3.4}$$

Here $f(x)$ is 2π-periodic in all space dimensions and we are interested in solutions which have the same property. We normalize the symbol $P(i\omega)$ of this problem

$$P(i\omega) = i\sum_{\nu=1}^{d} A_\nu\omega_\nu = |\omega| P(i\omega'),$$

with

$$|\omega|^2 = \sum |\omega_j|^2 \quad \text{and} \quad \omega' = \frac{\omega}{|\omega|}.$$

The following conditions 1. and 2. are necessary and sufficient for the problem (1.3.4) to be well posed:

1. For all $\omega' \in \mathbb{R}^d$, $|\omega'| = 1$, all eigenvalues of $P(i\omega')$ are purely imaginary
2. $P(i\omega')$ has a complete set of eigenvectors, which are uniformly linearly independent, i.e. there is a constant K and for every ω' a transformation $T(\omega')$ with
$$|T(\omega')| + |T^{-1}(\omega')| \leq K, \tag{1.3.5}$$
such that
$$T(\omega')\,P(i\omega')\,T^{-1}(\omega') = \begin{pmatrix} \lambda_1 & & 0 \\ & \ddots & \\ 0 & & \lambda_n \end{pmatrix}. \tag{1.3.6}$$

Definition 1.3.1 *The first order equation (1.3.4) is called weakly hyperbolic, if it only satisfies condition 1. If it satisfies conditions 1. and 2. it is called strongly hyperbolic. The problem (1.3.4) is called strictly hyperbolic, if for all $\omega \in \mathbb{R}^d$, $\omega \neq 0$, all eigenvalues of $P(i\omega)$ are purely imaginary and distinct.*

For symmetric hyperbolic problems (i.e. all matrices A_ν satisfy $A_\nu = A_\nu^*$), conditions 1. and 2. are naturally satisfied. For nonsymmetric systems, one can prove the following theorem:

Theorem 1.3.2 *The problem (1.3.4) is strongly hyperbolic, if the eigenvalues of $P(i\omega')$ are purely imaginary and the eigenvalues have constant multiplicity.*

In multiple space dimensions we consider, for every fixed $\omega = (\omega_1, \ldots, \omega_d)$,
$$\hat{u}_t(\omega, t) = |\omega| P(i\omega')\,\hat{u}(\omega, t),$$
and construct $\hat{H}(\omega') = T^*(\omega')\,T(\omega')$, such that
$$\hat{H}(\omega')\,P(i\omega') + P^*(i\omega')\,\hat{H}(\omega') = 0. \tag{1.3.7}$$
Let (u, v) again be the scalar product in L_2, we then define a new scalar product by
$$(u, v)_H = (u, Hv),$$
where
$$Hv(x, t) = \sum_\omega e^{i\langle\omega, x\rangle}\,\hat{H}(\omega')\,\hat{v}(\omega, t)$$
is well defined. It then follows from Parseval's relation (1.1.5) that
$$(u, Hu) = \sum_\omega \langle \hat{u}(\omega, t), \hat{H}(\omega')\,\hat{u}(\omega, t)\rangle.$$

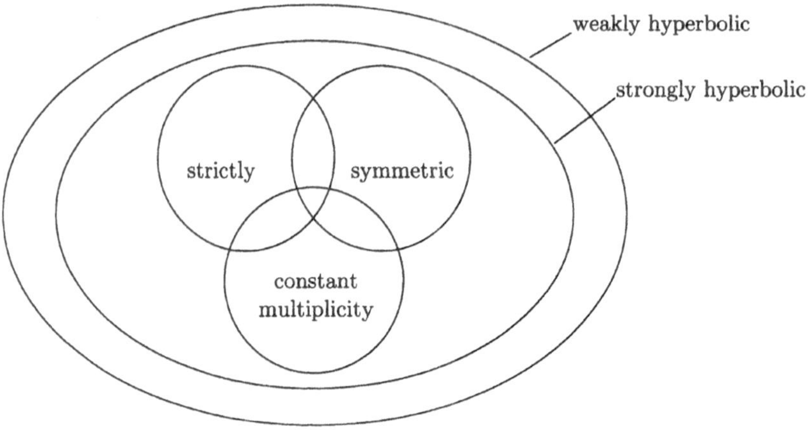

Figure 1.1:

Since by (1.3.5) and $\hat{H}(\omega') = T^*(\omega')\,T(\omega')$,

$$\begin{aligned} K^{-2}|\hat{u}|^2 &\leq \frac{1}{|T^{-1}|^2}|\hat{u}|^2 \\ &\leq \langle \hat{u}, \hat{H}(\omega')\,\hat{u}\rangle \\ &= |T\hat{u}|^2 \\ &\leq |T|^2|\hat{u}|^2 \\ &\leq K^2|\hat{u}|^2, \end{aligned}$$

we have

$$K^{-2}|\hat{u}|^2 \leq \langle \hat{u}, \hat{H}(\omega')\,\hat{u}\rangle \leq K^2|\hat{u}|^2.$$

Thus

$$\begin{aligned} K^{-2}\|u\|^2 &= K^{-2}\sum_{\omega}|\hat{u}|^2 \\ &\leq (u, Hu) \\ &= \sum_{\omega}\langle \hat{u}, \hat{H}\hat{u}\rangle \\ &\leq K^2\sum_{\omega}|\hat{u}|^2 \\ &= K^2\|u\|^2 \end{aligned}$$

and (u, Hv) defines a scalar product which is equivalent with the usual L_2-scalar product. Also (1.3.7) implies

$$\begin{aligned}(u, HPu) + (Pu, Hu) &= \sum_{\omega} \langle \hat{u}, \hat{H}(\omega') P(i\omega') \hat{u} \rangle + \sum_{\omega} \langle P(i\omega') \hat{u}, \hat{H}(\omega') \hat{u} \rangle \\ &= \sum_{\omega} \langle \hat{u}, \hat{H}(\omega') P(i\omega') \hat{u} \rangle + \sum_{\omega} \langle \hat{u}, P^*(i\omega') \hat{H}(\omega') \hat{u} \rangle \\ &= \sum_{\omega} \langle \hat{u}, \left(\hat{H}(\omega') P(i\omega') + P^*(i\omega') \hat{H}(\omega') \right) \hat{u} \rangle \\ &= 0.\end{aligned}$$

So we have shown that the new norm stays invariant:

$$\frac{\partial}{\partial t}(u, Hu) = 0.$$

1.4 General Systems with Constant Coefficients

We consider the general system

$$\begin{aligned} u_t(x,t) &= P(\partial/\partial x)\, u(x,t), \quad x \in \mathbb{R}^d, \quad t \geq 0, \\ u(x,0) &= f(x), \qquad\qquad\qquad x \in \mathbb{R}^d, \end{aligned} \tag{1.4.1}$$

where $P(\partial/\partial x)$ is a general differential operator of order m and $f(x)$ is again a smooth and 2π-periodic function. We denote by

$$\hat{u}_t(\omega, t) = P(i\omega)\, \hat{u}(\omega, t) \tag{1.4.2}$$

the Fourier transform of Equation (1.4.1).

Theorem 1.4.1 (The Petrovskii or eigenvalue condi-tion.) *A neccssary condition for well-posedness of (1.4.1) is that the eigenvalues $\lambda(\omega)$ of $P(i\omega)$ satisfy the inequality*

$$\operatorname{Re} \lambda(\omega) \leq \alpha \tag{1.4.3}$$

for a suitable $\alpha > 0$ independent of ω.

Proof. Let $\lambda(\omega)$ be any eigenvalue of $P(i\omega)$. Choose $\hat{u}(\omega, 0)$ as the corresponding eigenvector. Then

$$\hat{u}(\omega, t) = e^{\lambda(\omega)t} \hat{u}(\omega, 0).$$

Thus, if there is an eigenvalue $\lambda(\omega)$ which is not bounded from above, the problem cannot be well posed. □

Petrovskii defined well-posedness in such a way that the Petrovskii condition is also sufficient.

Definition 1.4.1 *We call a problem which satisfies the estimate*

$$\left|e^{P(i\omega)t}\right| \leq Ke^{\alpha t}\left(\left(|\omega|+1\right)t\right)^{p}$$

with K, α, p independent of ω weakly well posed.

The eigenvalue condition is sufficient for weakly well-posedness. The disadvantage is that weakly well-posedness is not stable against lower order perturbations, which in some sort reflect variable coefficients, as will be shown later.

Lemma 1.4.1 *The Petrovskii condition is sufficient for weakly well-posedness.*

In order to prove Lemma 1.4.1, we need

Lemma 1.4.2 (Schur's Lemma) *For every matrix A, there is a unitary matrix U such that U^*AU is upper triangular:*

$$U^*AU = \begin{pmatrix} \lambda_1 & a_{12} & \cdots & a_{1n} \\ 0 & \ddots & \ddots & \vdots \\ \vdots & \ddots & \ddots & a_{n-1,n} \\ 0 & \cdots & 0 & \lambda_n \end{pmatrix}. \tag{1.4.4}$$

Also the eigenvalues λ_j can be ordered arbitrarily for example such that

$$\operatorname{Re}\lambda_1 \geq \operatorname{Re}\lambda_2 \geq \ldots \geq \operatorname{Re}\lambda_n. \tag{1.4.5}$$

Proof of Lemma 1.4.1. Due to Schur's Lemma, there exists a unitary transformation U, $UU^* = I$, which transforms the symbol $P(i\omega)$ in the form (1.4.4). We make the substitution $\tilde{u} = U^*\hat{u}$ and obtain

$$\tilde{u}_t(\omega,t) = (\Lambda + A)\,\tilde{u}(\omega,t),$$

where $\Lambda = diag(\lambda_1,\ldots,\lambda_n)$ and A denotes the remaining upper triangular matrix. The change of variables

$$\tilde{u}(\omega,t) = e^{\Lambda t}v(\omega,t)$$

leads to the differential equation

$$\begin{aligned} v_t(\omega,t) &= e^{-\Lambda t}Ae^{\Lambda t}v(\omega,t) \\ &= \begin{pmatrix} 0 & a_{12}e^{(\lambda_2-\lambda_1)t} & \cdots & a_{1n}e^{(\lambda_n-\lambda_1)t} \\ \vdots & \ddots & \ddots & \vdots \\ \vdots & & \ddots & a_{n-1,n}e^{(\lambda_n-\lambda_{n-1})t} \\ 0 & \cdots & \cdots & 0 \end{pmatrix} v(\omega,t), \end{aligned}$$

which can be solved explicitly. The solution consists of polynomials in t multiplied by exponential functions with nonpositive real part. Since $P(\partial/\partial x)$ is a differential operator of order m, the symbol $P(i\omega)$ is a polynomial of order m in ω. Thus

$$|a_{ij}| \le K(|\omega|+1)^m. \tag{1.4.6}$$

Since we have ordered the λ_j in such a way that $\mathrm{Re}(\lambda_j - \lambda_i) < 0$ for $j > i$, it follows from (1.4.6) that all the elements of $e^{-\Lambda t} A e^{\Lambda t}$ are bounded terms of order $\mathcal{O}(|\omega|^m)$. Thus v is bounded by

$$|v(t)| \le K(|\omega|+1)^{(n-1)m}(t+1)^{n-1}. \tag{1.4.7}$$

It follows that

$$|\hat{u}| = |\tilde{u}| \le |e^{\Lambda t}||v| \le e^{\alpha t}|v|.$$

Thus the problem is weakly well posed. □

Theorem 1.4.2 (necessary and sufficient condition for well-posedness) *The problem (1.4.1) is well posed, if and only if,*

1. *the eigenvalue condition (1.4.3) is satisfied,*
2. *for every ω there exists a transformation T such that*

$$TP(i\omega)T^{-1} = \begin{pmatrix} \lambda_1 & a_{12} & \cdots & a_{1n} \\ 0 & \ddots & & \vdots \\ \vdots & \ddots & \ddots & a_{n-1,n} \\ 0 & \dots & 0 & \lambda_n \end{pmatrix} + D$$

with

(a) $|D| \le K_1$ (D is a bounded matrix),

(b) $|a_{ij}| \le K_2 \min(|\mathrm{Re}\,\lambda_i|, |\mathrm{Re}\,\lambda_j|)$ (the diagonal must control the off-diagonal),

(c) $|T| + |T^{-1}| \le K_3$,

where the constants K_1, K_2 and K_3 are independent of ω.

For the proof we refer to Kreiss and Oliger. To illustrate the theorem we consider the following special cases:

Hyperbolic systems: Since by our definition of hyperbolicity $\mathrm{Re}\,\lambda_j = 0$ it follows from condition (b) that the problem is well posed, if $a_{ij} = 0$, for all i, j. However, this is equivalent to our previous condition for well-posedness of hyperbolic problems.

1.4.1 Parabolic Systems

We first consider the special case

$$u_t(x,t) = P_m(\partial/\partial x)\, u(x,t) \tag{1.4.8}$$

where m is even and P_m is a homogeneous differential operator of order m, i.e.:

$$P_m(\partial/\partial x) = \sum_{|\nu|=m} A_\nu \frac{\partial^{|\nu|}}{\partial x_1^{\nu_1}\cdots\partial x_d^{\nu_d}}.$$

As before ν is a multi-index of order $|\nu| = \nu_1 + \cdots + \nu_d$. For $m = 2$, a simple example is

$$u_t(x,t) = au_{xx}(x,t) + bu_{xy}(x,t) + cu_{yy}(x,t).$$

Definition 1.4.2 *The system (1.4.8) is called parabolic if for all ω the eigenvalues $\lambda_j(\omega)$, $j = 1,\dots,n$, of $P_m(i\omega)$ satisfy*

$$\operatorname{Re}\lambda_j(\omega) \le -\delta\,|\omega|^m, \quad j = 1,\dots,n, \tag{1.4.9}$$

with some $\delta > 0$ independent of ω.

For our example, this condition is satisfied, if

$$-\left(a\,\omega_1^2 + b\,\omega_1\omega_2 + c\,\omega_2^2\right) \le -\delta|\omega|^2.$$

Theorem 1.4.3 *The Cauchy problem for the parabolic system (1.4.8) is well posed.*

Proof. Consider the operator $P_m(\partial/\partial x)$ with the symbol $P_m(i\omega)$. Transform $P_m(i\omega)$ to upper-triangular form by a unitary matrix $U = U(\omega)$:

$$UP_m(i\omega)U^* = \begin{pmatrix} \lambda_1 & a_{12} & \cdots & a_{1n} \\ & \ddots & & \vdots \\ & & \ddots & a_{n-1,n} \\ 0 & & & \lambda_n \end{pmatrix} = |\omega|^m \begin{pmatrix} \lambda_1' & a_{12}' & \cdots & a_{1n}' \\ & \ddots & & \vdots \\ & & \ddots & a_{n-1,n}' \\ 0 & & & \lambda_n' \end{pmatrix}.$$

The parabolicity assumption yields

$$\operatorname{Re}\lambda_j' \le -\delta, \quad j = 1,\dots,n.$$

The off-diagonal entries a_{ij}' are uniformly bounded by a constant K_1 independent of ω. By means of diagonal scaling, i.e. multiplication with

$$S = \begin{pmatrix} 1 & 0 & \cdots & 0 \\ 0 & s & & \vdots \\ \vdots & & \ddots & 0 \\ 0 & \cdots & 0 & s^{n-1} \end{pmatrix},$$

respectively S^{-1} we obtain

$$SUP_m(i\omega)U^*S^{-1} = |\omega|^m \begin{pmatrix} \lambda_1' & s^{-1}a_{12}' & \cdots & s^{-(n-1)}a_{1n}' \\ 0 & \ddots & \ddots & \vdots \\ \vdots & \ddots & \ddots & s^{-(n-1)}a_{(n-1),n}' \\ 0 & \cdots & 0 & \lambda_n' \end{pmatrix}.$$

We can assure that the influence of the off-diagonal entries is arbitrarily small by choosing s sufficiently large. Thus the entries of the diagonal dominate the off-diagonal entries and the problem is well posed according to Theorem 1.4.2. □

Theorem 1.4.4 *If the problem (1.4.8) is parabolic and thus well posed, we can find for any wave number ω a positive definite Hermitian matrix $\hat{H}$ such that*

$$\hat{H}P(i\omega) + P(i\omega)^*\hat{H} \le 2\tilde{\alpha}\hat{H},$$

$$K^{-1}I \le \hat{H} \le KI,$$

where the constants $\tilde{\alpha}$, K do not depend on ω.

Proof. By proceeding in the same way as in the proof of Theorem 1.4.3, we obtain again

$$SUP_m(i\omega)U^*S^{-1} = |\omega|^m \begin{pmatrix} \lambda_1' & s^{-1}a_{12}' & \cdots & s^{-(n-1)}a_{1n}' \\ 0 & \ddots & \ddots & \vdots \\ \vdots & \ddots & \ddots & s^{-(n-1)}a_{(n-1),n}' \\ 0 & \cdots & 0 & \lambda_n' \end{pmatrix} = B.$$

We choose $\hat{H} = (SU)^* SU$ and consider $\hat{H}P(i\omega) + P(i\omega)^*\hat{H}$:

$$\begin{aligned} &(SU)^* SUP_m(i\omega) + P_m(i\omega)^*(SU)^* SU \\ &= (SU)^*\left[SUP_m(i\omega)(SU)^{-1} + ((SU)^*)^{-1}P_m(i\omega)^*(SU)^*\right]SU \\ &= (SU)^*[B + B^*]SU. \end{aligned}$$

We constructed B in such a way that $B + B^*$ is symmetric, negative definite and bounded from above by $-\delta|\omega|^m$. Thus it follows that

$$\hat{H}P_m(i\omega) + P_m(i\omega)^*\hat{H} \le -\delta|\omega|^m\hat{H}.$$

The choice $\tilde{\alpha} = -\frac{1}{2}\delta|\omega|^m$ yields the desired estimate. □

By adding lower order terms to Equation (1.4.8) we obtain the following problem

$$u_t(x,t) = P_m(\partial/\partial x)\,u(x,t) + Q^{(m-1)}(\partial/\partial x)\,u(x,t) \tag{1.4.10}$$

where

$$Q^{(m-1)}(\partial/\partial x) = \sum_{|\nu|\le m-1} A_\nu \frac{\partial^{|\nu|}}{\partial x_1^{\nu_1}\cdots\partial x_d^{\nu_d}}.$$

We want to prove the following theorem:

Theorem 1.4.5 *The initial value problem (1.4.10) is well posed, if the principle part $P_m(\partial/\partial x)$ is parabolic.*

Proof. Consider the symbol of (1.4.10)

$$\hat{u}_t(\omega,t) = P_m(i\omega)\,\hat{u}(\omega,t) + Q^{(m-1)}(i\omega)\,\hat{u}(\omega,t).$$

With $\hat{H}$ given by the previous theorem for the principle part we conclude

$$\begin{aligned}
\frac{\partial}{\partial t}\langle\hat{u},\hat{H}\hat{u}\rangle &= \langle\hat{u}_t,\hat{H}\hat{u}\rangle + \langle\hat{u},\hat{H}\hat{u}_t\rangle \\
&= \langle\hat{u},\left(\hat{H}P_m + P_m^*\hat{H}\right)\hat{u}\rangle + \langle\hat{u},\left(\hat{H}Q^{(m-1)} + \left(Q^{(m-1)}\right)^*\hat{H}\right)\hat{u}\rangle \\
&\le -\delta|\omega|^m\langle\hat{u},\hat{H}\hat{u}\rangle + \tilde{K}\left(|\omega|^{m-1}+1\right)\langle\hat{u},\hat{H}\hat{u}\rangle \\
&\le \tilde{\alpha}\langle\hat{u},\hat{H}\hat{u}\rangle.
\end{aligned}$$

Here $\tilde{\alpha}$ is a sufficiently large constant independent of ω. □

1.4.2 Mixed Systems

We start out with the simple decoupled problem

$$\begin{aligned}
u_t(x,t) &= P_2(\partial/\partial x)\,u(x,t) \\
v_t(x,t) &= P_1(\partial/\partial x)\,v(x,t),
\end{aligned} \tag{1.4.11}$$

where $P_2(\partial/\partial x)$ is a second order parabolic operator and $P_1(\partial/\partial x)$ is a first order strongly hyperbolic operator. The problem (1.4.11) can be coupled by

$$\begin{aligned}
u_t(x,t) &= P_2(\partial/\partial x)\,u(x,t) + Q_{11}^{(1)}u(x,t) + Q_{12}^{(1)}v(x,t) \\
v_t(x,t) &= P_1(\partial/\partial x)\,v(x,t) + Q_{21}^{(1)}u(x,t) + Q_{22}^{(0)}v(x,t).
\end{aligned} \tag{1.4.12}$$

The operators $Q^{(1)}$ are of the form

$$Q^{(1)} = Q^{(1)}\left(\partial/\partial x\right) = \sum_{j=1}^{d} A_j \frac{\partial}{\partial x_j} + B,$$

and $Q_{22}^{(0)}$ is a zero order operator (matrix).

Theorem 1.4.6 *The Cauchy problem for (1.4.12) is well posed.*

Proof. We consider the Fourier transform of the system (1.4.12)

$$\begin{aligned} \hat{u}_t(\omega,t) &= \left(P_2(i\omega) + Q_{11}^{(1)}(i\omega)\right)\hat{u}(\omega,t) &+& \quad Q_{12}^{(1)}(i\omega)\,\hat{v}(\omega,t) \\ \hat{v}_t(\omega,t) &= \left(P_1(i\omega) + Q_{22}^{(0)}(i\omega)\right)\hat{v}(\omega,t) &+& \quad Q_{21}^{(1)}(i\omega)\,\hat{u}(\omega,t). \end{aligned} \tag{1.4.13}$$

We denote the norm related to $P_i(i\omega)$ by $\hat{H}_i$, $i = 1, 2$, and want to find an estimate for

$$\frac{\partial}{\partial t}\left(\langle \hat{u}, \hat{H}_2 \hat{u}\rangle + \langle \hat{v}, \hat{H}_1 \hat{v}\rangle\right). \tag{1.4.14}$$

Because P_2 is parabolic and P_1 is strictly hyperbolic, we have for suitable constants $\delta > 0$, $\tilde{K} > 0$ and for all ω

$$\begin{aligned} \hat{H}_2(\omega) P_2(i\omega) &+ P_2^*(i\omega)\hat{H}_2(\omega) &\le& -\delta|\omega|^2 \hat{H}, \\ \hat{H}_1(\omega) P_1(i\omega) &+ P_1^*(i\omega)\hat{H}_1(\omega) &=& \ 0, \\ \tilde{K}^{-1} I &\le \hat{H}_1(\omega), \hat{H}_2(\omega) &\le& \ \tilde{K} I. \end{aligned} \tag{1.4.15}$$

We look now at

$$\frac{\partial}{\partial t}\left(\langle \hat{u}, \hat{H}_2 \hat{u}\rangle + \langle \hat{v}, \hat{H}_1 \hat{v}\rangle\right) = \langle \hat{u}_t, \hat{H}_2 \hat{u}\rangle + \langle \hat{u}, \hat{H}_2 \hat{u}_t\rangle + \langle \hat{v}_t, \hat{H}_1 \hat{v}\rangle + \langle \hat{v}, \hat{H}_1 \hat{v}_t\rangle.$$

Substituting (1.4.13) into the above equation and using the properties (1.4.15) yields

$$\begin{aligned} &\frac{\partial}{\partial t}\left(\left\langle \hat{u}, \hat{H}_2 \hat{u}\right\rangle + \left\langle \hat{v}, \hat{H}_1 \hat{v}\right\rangle\right) \\ &= \left\langle P_2\hat{u} + Q_{11}^{(1)}\hat{u} + Q_{12}^{(1)}\hat{v}, \hat{H}_2\hat{u}\right\rangle + \left\langle \hat{u}, \hat{H}_2 P_2 \hat{u} + \hat{H}_2 Q_{11}^{(1)}\hat{u} + \hat{H}_2 Q_{12}^{(1)}\hat{v}\right\rangle \\ &\quad + \left\langle P_1\hat{v} + Q_{22}^{(0)}\hat{v} + Q_{21}^{(1)}\hat{u}, \hat{H}_1\hat{v}\right\rangle + \left\langle \hat{v}, \hat{H}_1 P_1 \hat{v} + \hat{H}_1 Q_{22}^{(0)}\hat{v} + \hat{H}_1 Q_{21}^{(1)}\hat{u}\right\rangle \\ &= \left\langle \hat{u}, \left(\hat{H}_2 P_2 + P_2^* \hat{H}_2\right)\hat{u}\right\rangle + \left\langle \hat{v}, \left(\hat{H}_1 P_1 + P_1^* \hat{H}_1\right)\hat{v}\right\rangle \\ &\quad + \left\langle Q_{11}^{(1)}\hat{u}, \hat{H}_2\hat{u}\right\rangle + \left\langle \hat{u}, \hat{H}_2 Q_{11}^{(1)}\hat{u}\right\rangle + \left\langle Q_{12}^{(1)}\hat{v}, \hat{H}_2\hat{u}\right\rangle + \left\langle \hat{u}, \hat{H}_2 Q_{12}^{(1)}\hat{v}\right\rangle \\ &\quad + \left\langle Q_{21}^{(1)}\hat{u}, \hat{H}_1\hat{v}\right\rangle + \left\langle \hat{v}, \hat{H}_1 Q_{21}^{(1)}\hat{u}\right\rangle + \left\langle Q_{22}^{(0)}\hat{v}, \hat{H}_1\hat{v}\right\rangle + \left\langle \hat{v}, \hat{H}_1 Q_{22}^{(0)}\hat{v}\right\rangle \end{aligned}$$

$$\begin{aligned}\le\ & -\delta|\omega|^2\left\langle\hat u,\hat H_2\hat u\right\rangle+K(|\omega|+1)\left\langle\hat u,\hat H_2\hat u\right\rangle\\ &+K(|\omega|+1)\left(\left\langle\hat v,\hat H_2\hat u\right\rangle+\left\langle\hat u,\hat H_2\hat v\right\rangle\right)\\ &+K(|\omega|+1)\left(\left\langle\hat u,\hat H_1\hat v\right\rangle+\left\langle\hat v,\hat H_1\hat u\right\rangle\right)+2K\left\langle\hat v,\hat H_1\hat v\right\rangle\end{aligned}$$

for some constant $K>0$. By using Schwarz's inequality

$$\left\langle\hat v,\hat H_2\hat u\right\rangle+\left\langle\hat u,\hat H_2\hat v\right\rangle\le 2|\hat v|_{\hat H_2}|\hat u|_{\hat H_2}$$

we obtain

$$\begin{aligned}&\frac{\partial}{\partial t}\left(\left\langle\hat u,\hat H_2\hat u\right\rangle+\left\langle\hat v,\hat H_1\hat v\right\rangle\right)\\ \le\ & -\delta|\omega|^2|\hat u|^2_{\hat H_2}+K|\omega|\left[|\hat u|^2_{\hat H_2}+2|\hat v|_{\hat H_2}|\hat u|_{\hat H_2}+2|\hat v|_{\hat H_1}|\hat u|_{\hat H_1}\right]\\ &+K\left[|\hat u|^2_{\hat H_2}+2|\hat v|_{\hat H_2}|\hat u|_{\hat H_2}+2|\hat v|_{\hat H_1}|\hat u|_{\hat H_1}+|\hat v|_{\hat H_1}\right].\end{aligned}$$

For any two real numbers a,b we know that $2ab\le a^2+b^2$, and therefore

$$2\varepsilon|\omega|\left\langle\hat u,\hat H_2\hat u\right\rangle^{1/2}\frac{1}{\varepsilon}\left\langle\hat v,\hat H_1\hat v\right\rangle^{1/2}\le\varepsilon^2|\omega|^2\left\langle\hat u,\hat H_2\hat u\right\rangle+\frac{1}{\varepsilon^2}\left\langle\hat v,\hat H_1\hat v\right\rangle.$$

By using this inequality for ε sufficiently small, we force the term which is still depending on ω to stay sufficiently small. We call this technique *weighted squaring*:

$$\begin{aligned}\frac{\partial}{\partial t}\left(\langle\hat u,\hat H_2\hat u\rangle+\langle\hat v,\hat H_1\hat v\rangle\right)\ &\le\ \left(-(\delta/4)|\omega|^2+K_1(|\omega|+1)\right)\langle\hat u,\hat H_2\hat u\rangle\\ &\quad+K_2\langle\hat v,\hat H_1\hat v\rangle\\ &\le\ 2\tilde\alpha\left(\langle\hat u,\hat H_2\hat u\rangle+\langle\hat v,\hat H_1\hat v\rangle\right).\end{aligned}$$

The theorem then follows from Gronwall's lemma.

As an example we consider the compressible Navier-Stokes equations. Linearized around a constant state they are of the above form:

$$\begin{aligned}u_t &+ Uu_x + Vu_y + p_x &&= \nu\Delta u\\ v_t &+ Uv_x + Vv_y + p_y &&= \nu\Delta v\\ p_t &+ Up_x + Vp_y + u_x+v_y &&= 0,\end{aligned}$$

where $\nu>0$ stands for the viscosity. Since first order derivatives with respect to x and y do not affect the parabolicity and since first order derivatives of u

and v do not affect the hyperbolicity, we can drop them and obtain

$$\begin{aligned} u_t &= \nu \Delta u \\ v_t &= \nu \Delta v \\ p_t + U p_x + V p_y &= 0. \end{aligned} \tag{1.4.16}$$

We state a more general version of Theorem 1.4.4.

Theorem 1.4.7 *If the initial value problem (1.4.1) is well posed, we can find for any wave number ω a positive definite Hermitian matrix $\hat{H}$ such that*

$$\hat{H} P(i\omega) + P(i\omega)^* \hat{H} \leq 2\alpha \hat{H},$$

$$K^{-1} I \leq \hat{H} \leq K I,$$

where the constants α, K do not depend on ω.

For the proof we again refer to Kreiss and Oliger.

1.5 Linear Systems with Variable Coefficients

In this chapter we have only treated systems with constant coefficients. This might look very restrictive and irrelevant because the differential equations arising in applications are typically nonlinear. As we shall see in the last chapter the existence of solutions of nonlinear problems can be reduced to the existence of solutions of linear problems with variable coefficients. For hyperbolic, parabolic and mixed hyperbolic-parabolic systems the "frozen coefficient test" is the guiding principle. We shall explain this for first order systems.

Consider the Cauchy problem

$$u_t(x,t) = P(x,t,\partial/\partial x)\, u(x,t), \tag{1.5.1}$$

for a first order system. We consider all systems with constant coefficients

$$v_t(x,t) = P(x_0,t_0,\partial/\partial x)\, v(x,t), \tag{1.5.2}$$

where the coefficients are frozen at an arbitrary point (x_0, t_0). If

1. all constant coefficient systems (1.5.2) are symmetric hyperbolic or
2. all constant coefficient systems (1.5.2) are strongly hyperbolic and the multiplicity of the eigenvalues of the symbol does not depend on x, t and ω,

then the problem (1.5.1) is well posed.

1.6 Remarks

Detailed duscussions of the material in Chapter 1 are contained in Chapter 2, 3 and 6 of Kreiss and Lorenz and in Chapter 4 of Gustafsson, Kreiss ans Oliger.

Chapter 2

Half Plane Problems

2.1 Hyperbolic Systems in One Dimension

In this section, we discuss the concept of well-posedness for half plane problems. Consider a strongly hyperbolic system

$$\begin{aligned} u_t(x,t) &= Au_x(x,t) + F(x,t) \quad \text{for} \quad x \geq 0, \quad t \geq 0, \\ u(x,0) &= f(x) \qquad\qquad\qquad\quad \text{for} \quad x \geq 0, \end{aligned} \tag{2.1.1}$$

We assume that A is a constant nonsingular diagonal matrix and use the notation $u = (u_1, \ldots, u_n)^T$. We write A in the form

$$A = \begin{pmatrix} -\Lambda_1 & 0 \\ 0 & \Lambda_2 \end{pmatrix},$$

such that $\Lambda_1 = diag(\lambda_1, \cdots, \lambda_r) > 0$ and $\Lambda_2 = diag(\lambda_{r+1}, \cdots, \lambda_n) > 0$. The initial data $f(x)$ and the forcing function $F(x,t)$ are smooth functions, which belong to L_2 , i.e.

$$\|f\|_{L^2}^2 = \int_0^\infty |f|^2 \, dx < \infty, \quad \|F(\cdot,t)\|_{L^2}^2 < \infty,$$

for every fixed t. In addition, we give boundary conditions

$$Lu(0,t) = g(t), \tag{2.1.2}$$

where L stands for q linearly independent relations

$$L = \begin{pmatrix} l_{11} & \cdots & l_{1n} \\ \vdots & & \vdots \\ l_{q1} & \cdots & l_{qn} \end{pmatrix}.$$

We are only interested in solutions which belong to L_2.

By examining test problems, we can determine the number q of boundary conditions, such that the problem is well posed.

We consider (2.1.1) with $F \equiv 0$, $g \equiv 0$.

Lemma 2.1.1 *The problem is not well posed, if there exists a solution of the form*

$$u(x,t) = e^{st}\varphi(x),$$
$$L\varphi(0) = 0,$$

with $\|\varphi\|_{L^2} < \infty$ *and* $\operatorname{Re} s > 0$.

Proof. If $u(x,t) = e^{st}\varphi(x)$ is a solution, then

$$u_\alpha(x,t) = u(\alpha x, \alpha t) = e^{\alpha st}\varphi(\alpha x)$$

is also a solution with initial data $u(\alpha x, 0) = \varphi(\alpha x)$, $\alpha > 0$. Since α is an arbitrary parameter, we can construct a solution which grows arbitrarily fast. □

In order to obtain solutions of the form $u(x,t) = e^{st}\varphi(x)$, we substitute $u = e^{st}\varphi(x)$ into Equation (2.1.1) and obtain the eigenvalue problem

$$\begin{aligned} s\varphi(x) &= A\varphi_x(x), \\ L\varphi(x) &= 0, \quad \|\varphi\| < \infty. \end{aligned} \tag{2.1.3}$$

It follows that the problem (2.1.1) is not well posed, if (2.1.3) has an eigenvalue s with $\operatorname{Re} s > 0$. To discuss the eigenvalue problem (2.1.3) we make use of the block form of A and write

$$s\varphi^I(x) = -\Lambda_1 \varphi_x^I(x),$$

which is solved by

$$\varphi^I(x) = e^{-s\Lambda_1^{-1}x}\varphi^I(0).$$

Since the entries of Λ_1 are positive, $e^{-s\Lambda_1^{-1}x}$ decays exponentially, thus $\varphi^I(0)$ can be arbitrary. In analogy we obtain

$$\varphi^{II}(x) = e^{s\Lambda_2^{-1}x}\varphi^{II}(0).$$

Here the term $e^{s\Lambda_2^{-1}x}$ is exponentially increasing. Since we require that the solution $\varphi^{II}(x)$ is in L_2, it follows that

$$\varphi^{II}(0) = 0.$$

By sorting the variables, we rewrite the boundary conditions (2.1.2) in the following way

$$L\varphi(0) = D_1\varphi^I(0) + D_2\varphi^{II}(0) = 0. \tag{2.1.4}$$

Consider Equation (2.1.4) at time $t = 0$. With $\varphi^{II}(0) = 0$ we obtain

$$D_1\varphi^I(0) = 0.$$

We thus have proved the following lemma:

Lemma 2.1.2 *The eigenvalue problem (2.1.3) has an eigenvalue s with* $\operatorname{Re} s > 0$, *if and only if,*

$$D_1\varphi^I(0) = 0 \tag{2.1.5}$$

has a non-trivial solution.

Let again q denote the number of boundary conditions and r denote the number of negative eigenvalues of A. It follows that for

- $q < r$: Equation (2.1.5) has non-trivial solutions and thus the problem (2.1.1) is not well posed.
- $q = r$: the problem (2.1.1) is not well posed, if $\det|D_1| = 0$.
- $q > r$: Equation (2.1.5) is over-determined. Our problem has, in general, no solution.

To see why $q > r$ is unreasonable, we consider (2.1.1)–(2.1.2) with $f = 0$, $g = 0$ and solve the problem by applying the Laplace transform. We obtain

$$\int_0^\infty e^{-st}u_t(x,t)\,dt = A\int_0^\infty e^{-st}u_x(x,t)\,dt + \int_0^\infty e^{-st}F(x,t)\,dt. \tag{2.1.6}$$

We assume that $\operatorname{Re} s > 0$ is so large that $e^{-st}F(x,t)$, $e^{-st}u(x,t)$ and their derivatives decay exponentially. We denote by

$$\hat{u}(x,s) = \int_0^\infty e^{-st}u(x,t)\,dt$$

the Laplace transform of $u(x,t)$. Integration by parts shows that the Laplace transform $\hat{u}_t(x,t)$ of $u_t(x,t)$ is given by

$$\int_0^\infty e^{-st}u_t(x,t)\,dt = \underbrace{\left[e^{-st}u(x,t)\right]_0^\infty}_{=0} + s\int_0^\infty e^{-st}u(x,t)\,dt = s\hat{u}(x,s).$$

Thus

$$\hat{u}_t(x,s) = s\hat{u}(x,s).$$

Further we know that

$$\int_0^\infty e^{-st} u_x(x,t)\,dt = \frac{\partial}{\partial x}\int_0^\infty e^{-st} u(x,t)\,dt = \hat{u}_x(x,s).$$

Therefore (2.1.6) becomes

$$\begin{aligned} s\hat{u}(x,s) &= A\hat{u}_x(x,s) + \hat{F}, \\ L\hat{u}(0,s) &= 0, \\ \|\hat{u}(\cdot,s)\|^2 &< \infty. \end{aligned} \tag{2.1.7}$$

Because of the block structure of A we can rewrite (2.1.7)

$$\begin{aligned} \hat{u}_x^I(x,s) &= -s\Lambda_1^{-1}\hat{u}^I(x,s) + \Lambda_1^{-1}\hat{F}^I(x,s), & \Lambda_1 > 0, \\ \hat{u}_x^{II}(x,s) &= s\Lambda_2^{-1}\hat{u}^{II}(x,s) - \Lambda_2^{-1}\hat{F}^{II}(x,s), & \Lambda_2 > 0, \end{aligned} \tag{2.1.8}$$

with boundary conditions

$$L\hat{u}(0,s) = D_1\hat{u}^I(0,s) + D_2\hat{u}^{II}(0,s) = 0. \tag{2.1.9}$$

We obtain the general solution of (2.1.8) by using the variation of constants technique

$$\begin{aligned} \hat{u}^{II}(x,s) &= -\int_\infty^x e^{s\Lambda_2^{-1}(x-\sigma)}\Lambda_2^{-1}\hat{F}^{II}(\sigma,s)\,d\sigma, \\ \hat{u}^I(x,s) &= \int_0^x e^{-s\Lambda_1^{-1}(x-\sigma)}\Lambda_1^{-1}\hat{F}^I(\sigma,s)\,d\sigma + e^{-s\Lambda_1^{-1}x}\hat{u}^I(0,s). \end{aligned}$$

Thus $\hat{u}^{II}(0,s)$ is uniquely determined by the differential equation. For given $\hat{u}^{II}(0,s)$, $\hat{u}^I(0,s)$ is uniquely determined by (2.1.9) if the number of boundary conditions is equal to the number of negative eigenvalues and if $\det(D_1) \neq 0$. If the Laplace transformed equation has no solution, then the original problem has no solution either. Thus it is a necessary condition for well-posedness of Equation (2.1.1) with boundary conditions (2.1.2) that the number of boundary conditions is equal to the number of negative eigenvalues and that $\det(D_1) \neq 0$.

As we will discuss in Section 3.2, this technique can be generalized to difference equations, because we did not make use of the theory of characteristics.

By considering the characteristics for the one-dimensional problem we immediately get the same results as we will show by means of an example. Consider the one-dimensional problem

$$\begin{aligned} u_t(x,t) &= \lambda u_x(x,t), \quad x \geq 0, \quad t \geq 0, \\ u(x,0) &= f(x), \\ u(0,t) &= g(t). \end{aligned} \tag{2.1.10}$$

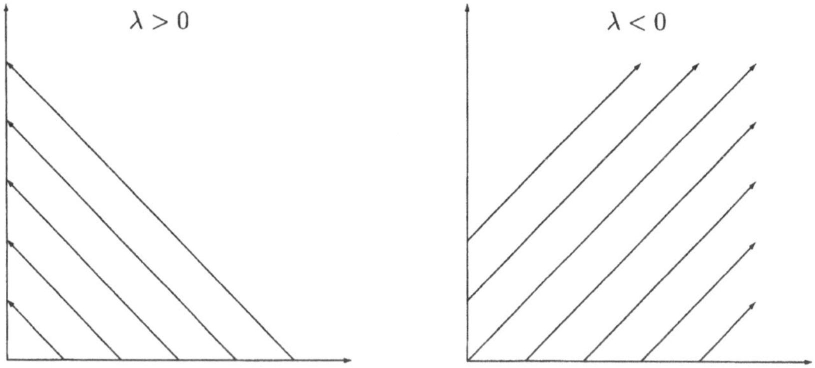

Figure 2.1: Characteristics of Equation (2.1.10) for $\lambda > 0$ and $\lambda < 0$

The solution of this problem is given by

$$u(x,t) = f(x + \lambda t),$$

and is thus constant along the lines $x + \lambda t = const.$

For $\lambda > 0$, the solution is fully determined by the initial data, no boundary conditions are needed. In the case $\lambda < 0$, we need boundary data at $x = 0$ to determine the solution.

2.2 Hyperbolic Systems in Two Dimensions

Consider the initial boundary value problem defined on the half plane $x \geq 0$, $-\infty < y < \infty$, for time $t \geq 0$

$$\begin{aligned} u_t(x,y,t) &= Au_x(x,y,t) + Bu_y(x,y,t), \\ u(x,y,0) &= f(x,y), \\ Lu(0,y,t) &= 0. \end{aligned} \tag{2.2.1}$$

Here f is a smooth function with compact support. We are interested in smooth solutions which belong to L_2 for every fixed t. We assume that the system is strongly hyperbolic and make a preliminary transformation of A such that

$$A = \begin{pmatrix} -\Lambda_1 & 0 \\ 0 & \Lambda_2 \end{pmatrix},$$

where $\Lambda_1 = diag(\lambda_1, \cdots, \lambda_r) > 0$ and $\Lambda_2 = diag(\lambda_{r+1}, \cdots, \lambda_n) > 0$. In general, B is not diagonal. We rewrite the boundary conditions

$$Lu(0, y, t) \equiv D_1 u^I(0, y, t) + D_2 u^{II}(0, y, t) = 0. \tag{2.2.2}$$

We apply the Laplace transform in time and the Fourier transform in y to system (2.2.1) and obtain

$$s\hat{u}(x, \omega_2, s) = A\hat{u}_x(x, \omega_2, s) + i\omega_2 B\hat{u}(x, \omega_2, s).$$

Again we look at test problems to derive necessary conditions for well-posedness.

Lemma 2.2.1 *The problem (2.2.1) is not well posed, if we can find one solution u of (2.2.1) of the type*

$$u(x, y, t) = e^{st + i\omega_2 y}\varphi(x),$$
$$L\varphi(0) = 0,$$

with $\|\varphi(\cdot)\|_{L_2} < \infty$ *and* $\operatorname{Re} s > 0$.

Proof. If $u(x, y, t)$ is a solution, then

$$u_\alpha(x, y, t) = e^{\alpha st + i\omega_2 \alpha y}\varphi(\alpha x), \quad \text{for } \alpha > 0,$$

is also a solution. Since α is arbitrary, we can construct solutions growing arbitrarily fast and thus the problem is not well posed. □

To find out when solutions of this type exist, we introduce the ansatz

$$u(x, y, t) = e^{st + i\omega_2 y}\varphi(x)$$

into Equation (2.2.1), and obtain for $\varphi(x)$

$$\begin{aligned} s\varphi(x) &= A\varphi_x(x) + i\omega_2 B\varphi(x), \\ L\varphi(0) &= 0, \quad \|\varphi\|_{L_2} < \infty. \end{aligned} \tag{2.2.3}$$

If we can find a solution of Equation (2.2.3) with $\operatorname{Re} s > 0$, then the problem (2.2.1) is not well posed. (2.2.3) is a system of ordinary differential equations which we can rewrite as

$$\varphi_x(x) = A^{-1}(sI - i\omega_2 B)\varphi(x) =: M\varphi(x),$$
$$L\varphi(0) = 0.$$

Lemma 2.2.2 *For* $\operatorname{Re} s > 0$, *the eigenvalues κ of M are never purely imaginary. There are exactly as many κ with positive (negative) real part as there are positive (negative) eigenvalues of A.*

Proof. By choosing the ansatz

$$\varphi(x) = e^{\kappa x}\varphi_0$$

for some κ with positive (negative) real part, we obtain that φ_0 must be a solution of the algebraic eigenvalue problem

$$(s - A\kappa - i\omega_2 B)\,\varphi_0 = 0.$$

Assume that $\kappa_1 = i\xi$, $\xi \in \mathbb{R}$. Then

$$(s - (Ai\xi + i\omega_2 B))\,\varphi_0 = 0.$$

Thus s must be an eigenvalue of the symbol $Ai\xi + i\omega_2 B$. By assumption the problem is strongly hyperbolic, thus the eigenvalues of the symbol are purely imaginary, which is a contradiction to our assumption that $\operatorname{Re} s > 0$. To show the second part of the lemma, we first consider the case $\omega_2 = 0$ for which the statement follows from our previous discussion. Since κ depends continuously on ω_2 and since the real part of κ cannot become zero, it has to stay positive (negative). □

Determine the general solution

$$\varphi(x) = \sum_{\operatorname{Re}\kappa_j > 0} e^{\kappa_j x} P_j.$$

If the multiplicity of κ_j is m_j, then P_j is a polynomial in x of order $m_j - 1$. The solution depends on r free parameters τ_j. Introducing the general solution into the boundary conditions leads to a homogeneous linear system for these parameters which we write as $C\tau = 0$. Thus the problem is not well posed if $\det(C) = 0$.

Example 1. Consider the following problem

$$\begin{pmatrix} u \\ v \end{pmatrix}_t = \begin{pmatrix} -1 & 0 \\ 0 & 1 \end{pmatrix} \begin{pmatrix} u \\ v \end{pmatrix}_x + \begin{pmatrix} 0 & 1 \\ 1 & 0 \end{pmatrix} \begin{pmatrix} u \\ v \end{pmatrix}_y \tag{2.2.4}$$

for $x \geq 0$, $-\infty < y < \infty$, $t \geq 0$.

Since the matrix

$$\begin{pmatrix} -1 & 0 \\ 0 & 1 \end{pmatrix}$$

has one negative eigenvalue, we know from the previous discussion that we need exactly one boundary condition

$$u\,(0, y, t) = \alpha v\,(0, y, t)\,. \tag{2.2.5}$$

To make the example more interesting, we allow α to be a complex number. We denote by $\underline{u}$ the vector $(u,v)^T$ and try to find a solution of the form

$$\begin{aligned} \underline{u}(x,y,t) &= e^{st+i\omega y}\begin{pmatrix} \varphi(x) \\ \psi(x) \end{pmatrix}, \\ \varphi(0) &= \alpha\psi(0), \end{aligned}$$

with $\|\underline{u}\| < \infty$. It follows from Lemma 2.2.1. that the problem (2.2.4) is not well posed if there is a non-trivial solution with $\operatorname{Re} s > 0$. Substituting the above expression into Equation (2.2.4) leads to the system of ordinary differential equations

$$s\underline{\varphi}(x) = \begin{pmatrix} -1 & 0 \\ 0 & 1 \end{pmatrix}\underline{\varphi}_x(x) + i\omega\begin{pmatrix} 0 & 1 \\ 1 & 0 \end{pmatrix}\underline{\varphi}(x), \tag{2.2.6}$$

where $\underline{\varphi}(x)$ denotes the vector $(\varphi(x), \psi(x))^T$. We can rewrite Equation (2.2.6) as follows

$$\underline{\varphi}_x(x) = M\underline{\varphi}(x) \tag{2.2.7}$$

with

$$M = \begin{pmatrix} -1 & 0 \\ 0 & 1 \end{pmatrix}\left(Is - i\omega\begin{pmatrix} 0 & 1 \\ 1 & 0 \end{pmatrix}\right) = \begin{pmatrix} -s & i\omega \\ -i\omega & s \end{pmatrix}.$$

The general solution of Equation (2.2.7) belonging to L_2 is

$$\varphi(x) = e^{\kappa_1 x}\underline{\varphi}_0$$

with

$$\kappa_1 = -\sqrt{s^2+\omega^2} \quad \text{and} \quad \underline{\varphi}_0 = \begin{pmatrix} s+\sqrt{s^2+\omega^2} \\ i\omega \end{pmatrix}.$$

The boundary condition is satisfied if

$$s + \sqrt{s^2+\omega^2} = \alpha i\omega,$$

which becomes with $s' = \dfrac{s}{|\omega|}$ and $\omega' = \dfrac{\omega}{|\omega|}$

$$y(s) := s' + \sqrt{s'^2+1} = \pm i\alpha. \tag{2.2.8}$$

$y(s)$ represents a mapping of the right half plane $\operatorname{Re} s > 0$ onto the set $|y| > 1$, $\operatorname{Re} y > 0$ (see Figure 2.2). Therefore (2.2.8) has a solution and the problem is ill posed, if $|\alpha| > 1$, $\operatorname{Im}\alpha \neq 0$. For real α and complex α with $|\alpha| \leq 1$, we do not get bad test solutions. (see Figure 2.3)

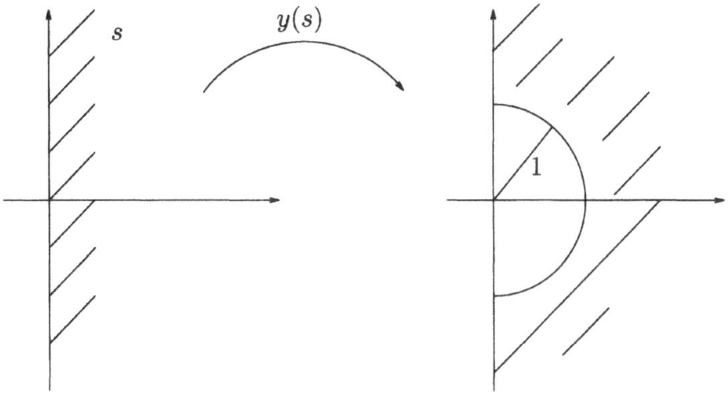

Figure 2.2:

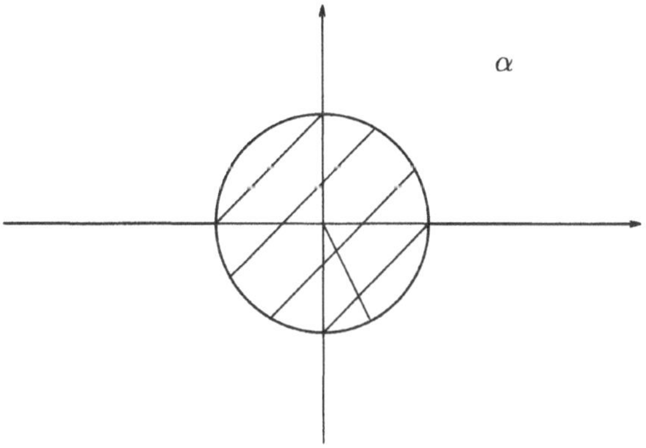

Figure 2.3:

Example 2. Consider the one-dimensional system

$$\begin{array}{rcrcl} u_t(x,t) & + & u_x(x,t) & = & 0 \\ v_t(x,t) & - & v_x(x,t) & = & 0, \end{array} \tag{2.2.9}$$

for $t \geq 0$ and $x \geq 0$ and with boundary conditions

$$u_x(0,t) = v(0,t), \quad \|u\|_{L_2} < \infty, \ \|v\|_{L_2} < \infty, \tag{2.2.10}$$

or

$$u(0,t) = v_x(0,t), \quad \|u\|_{L_2} < \infty, \ \|v\|_{L_2} < \infty. \tag{2.2.11}$$

The problem (2.2.9) is ill posed, if it has a non-trivial solution of the form

$$u(x,t) = e^{st}\varphi(x) \quad \text{and} \quad v(x,t) = e^{st}\psi(x), \tag{2.2.12}$$

where $\operatorname{Re} s > 0$. Substituting (2.2.12) into (2.2.9) leads to

$$\begin{array}{rcrcl} s\varphi(x) & + & \varphi_x(x) & = & 0 \\ s\psi(x) & - & \psi_x(x) & = & 0. \end{array} \tag{2.2.13}$$

Thus we have

$$\varphi(x) = e^{-sx}\varphi_0 \quad \text{and} \quad \psi(x) = e^{sx}\psi_0. \tag{2.2.14}$$

Because we require that $\psi(x)$ is square integrable, it follows that $\psi_0 = 0$. Thus $v(x,t) \equiv 0$. From the boundary condition (2.2.10) or (2.2.11), it follows that also $u(x,t) \equiv 0$. Thus there is no non-trivial solution of the form (2.2.12). The equations (2.2.9) with boundary conditions (2.2.10) or (2.2.11) have no bad test solution.

Example 3. We again consider Equation (2.2.9), but now for $0 \leq x \leq 1$. We replace the requirement of $u, v \in L_2$ by a boundary condition at $x = 1$. First we consider (2.2.9) together with the two boundary conditions

$$\begin{array}{rcl} u_x(0,t) & = & v(0,t), \\ v(1,t) & = & u(1,t). \end{array} \tag{2.2.15}$$

We obtain the same solutions (2.2.14) as in the above example, but now we have to determine the two free parameters φ_0 and ψ_0 such that the boundary conditions (2.2.15) are satisfied

$$\left.\begin{array}{rcl} -s\varphi_0 & = & \psi_0 \\ e^{-s}\varphi_0 & = & e^{s}\psi_0 \end{array}\right\} \Rightarrow \ e^{-2s}\varphi_0 = -s\varphi_0.$$

Thus s has to satisfy $e^{-2s} = -s$ and it follows

$$e^{-2\operatorname{Re} s} = |s|,$$

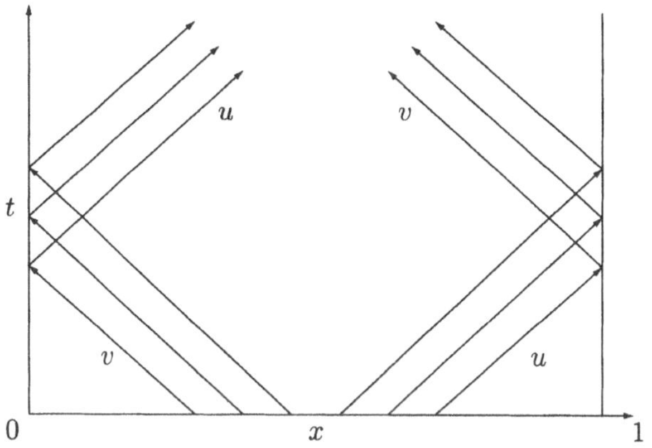

Figure 2.4:

which is not possible for $\operatorname{Re} s > 1$. Thus we cannot construct test solutions which grow arbitrarily fast.

Next we consider (2.2.9) with the boundary conditions

$$\begin{aligned} u(0,t) &= v_x(0,t), \\ v(1,t) &= u(1,t). \end{aligned} \qquad (2.2.16)$$

Proceeding in the same way as before we obtain the following condition for the parameter s

$$s = e^{2s}.$$

One can show that this equation has solutions with arbitrary large $\operatorname{Re} s$. Thus we can construct test solutions which grow arbitrarily fast.

The above results lead to the following conclusion: If the problem passes our test for half plane problems, we cannot be sure that it also passes this test if we add another boundary.

Geometric Interpretation. We first consider (2.2.9) with boundary conditions (2.2.15). Using the differential equation, we can replace $u_x(0,t)$ by $-u_t(0,t)$, i.e. we consider (2.2.9) with boundary conditions

$$\begin{aligned} u(0,t) &= \int_0^t v(0,\tau)\, d\tau, \\ v(1,t) &= u(1,t). \end{aligned} \qquad (2.2.17)$$

Along the characteristics $x + t = const.$, v is transported to the boundary $x = 0$. Then the boundary values of u are obtained by integrating v. u is then transported into the interior of the domain along the characteristics $t - x = const.$ Thus the reflection amounts to a smoothing process.

Correspondingly the boundary conditions (2.2.16) can be written as

$$\begin{aligned} u(0,t) &= -v_t(0,t), \\ u(1,t) &= v(1,t). \end{aligned} \tag{2.2.18}$$

thus the reflection amounts to a differentiation process and we loose one derivative at every reflection. Even if the initial condition is n times differentiable, after $(n+1)$ reflections at the boundary $x = 0$ the solution is not differentiable.

Example 4. Consider (2.2.4) in the domain $0 \le x \le 1$, $y \in \mathbb{R}$, with real boundary conditions

$$\begin{aligned} u(0,y,t) &= \alpha v(0,y,t), \qquad \alpha > 0, \\ u(1,y,t) &= \beta v(1,y,t), \qquad |\beta| < 1. \end{aligned} \tag{2.2.19}$$

We claim that this problem is not well posed. By Fourier transform in the y-direction, one can find arbitrarily fast growing solutions. The calculation is left to the reader.

Thus it is not true, that by counting the negative eigenvalues and matching the number of boundary conditions, one automatically gets well-posed problems.

2.3 Well-Posed Half Plane Problems

We consider a system of partial differential equations

$$u_t(x,y,t) = P(\partial/\partial x, \partial/\partial y)\, u(x,y,t) + F(x,y,t) \tag{2.3.1}$$

for $x \ge 0$, $y \in \mathbb{R}$. For example the system (2.2.1). At $t = 0$ we give initial data

$$u(x,y,0) = f(x,y)$$

and at $x = 0$ boundary conditions

$$Lu(0,y,t) = g(y,t), \quad \|u(\cdot,t)\|_{R_d} < \infty.$$

In general, L can also be a differential operator, but for first order systems, L is mostly of the form (2.2.2). We assume that F, f and g are smooth functions, which decay rapidly for $|x| \to \infty$, $|y| \to \infty$, and we are interested in solutions with the same property.

We can always assume that $g \equiv 0$. Otherwise, we construct a suitable function $\varphi(y,t)\,e^{-x}$ with $L\varphi = g$ and make a change of variables $v = u - \varphi$. Also, if we can solve the homogeneous system

$$\begin{aligned} v_t(x,y,t) &= P(\partial/\partial x, \partial/\partial y)\, v(x,y,t) \\ v(x,y,0) &= f(x,y) \\ Lv(0,y,t) &= 0, \quad \|v(\cdot,t)\|_{R_d} \le \infty, \end{aligned} \tag{2.3.2}$$

then we can also solve the inhomogeneous system (2.3.1) using Duhamel's principle. Therefore, as for the Cauchy problem, we need to define well-posedness only for the homogeneous system with zero boundary data (2.3.2).

Definition 2.3.1 *The problem (2.3.2) is well posed, if*

1. *For every smooth f, there is a solution and*
2. *there are constants K, α such that*

$$\|v(\cdot,t)\|_{R_d} \le K e^{\alpha t} \|f(\cdot)\|_{R_d}.$$

Here $\|\cdot\|_{R_d}$ denotes the L_2-norm for the domain $x \ge 0$, $y \in \mathbb{R}$.

This definition is adequate, if we can derive energy estimates by integration by parts. Consider for example (2.2.1), (2.2.2) and assume that $A = A^*$ and $B = B^*$ are constant Hermitian matrices. We need the following identities

$$\begin{aligned} (u, Bv_y)_{R_d} &= -(v_y, Bu)_{R_d}, \\ (u, Au_x)_{R_d} &= -(Au_x, u)_{R_d} - \langle u, Au\rangle|_{x=0}, \end{aligned}$$

which can be easily obtained by integration by parts. For the differential equation, we obtain

$$\begin{aligned} \frac{\partial}{\partial t}(u,u)_{R_d} &= (u, Au_x) + (Au_x, u) + (u, Bu_y) + (Bu, u_y) \\ &= -\langle u, Au\rangle|_{x=0}. \end{aligned}$$

Using the boundary conditions and the notation $S = D_1^{-1} D_2$ we have

$$\begin{aligned} -\langle u, Au\rangle|_{x=0} &= \left(u^I, \Lambda_1 u^I\right)\big|_{x=0} - \left(u^{II}, \Lambda_2 u^{II}\right) \\ &= -\langle u^{II}, (\Lambda_2 - S^*\Lambda_1 S)\, u^{II}\rangle \\ &\le 0 \quad \text{provided} \quad \Lambda_2 > S^*\Lambda_1 S. \end{aligned}$$

Therefore we obtain an energy estimate provided $|S|$ is sufficiently small.

There is an extensive literature on energy estimates.

One has formalized the process in the following way:

Definition 2.3.2 *We call the differential operator P together with the boundary condition a semibounded operator, if there is a constant α such the inequality*

$$(w, Pw)_{R_d} + (Pw, w)_{R_d} \leq 2\alpha \|w\|^2_{R_d} \tag{2.3.3}$$

holds for all smooth w which satisfy the boundary conditions.

If P is semibounded, then all smooth solutions satisfy

$$\frac{\partial}{\partial t}(u, u)_{R_d} = (u, Pu)_{R_d} + (Pu, u)_{R_d} \leq 2\alpha \|u\|^2_{R_d}. \tag{2.3.4}$$

The energy method is very powerful, when it works. However, if it fails, then we cannot conclude that the problem is ill posed.

As for the Cauchy problem, one has generalized the concept, requiring that (2.3.3) holds in a suitable norm. For the Cauchy problem we have shown that we always can find a norm in which the operator is semibounded. No such result is known for half plane problems.

In the next section, we weaken our concept of well-posedness such that our test is always a necessary and sufficient condition.

2.4 Well-Posed Problems in the Generalized Sense

We consider again the system

$$u_t(x, y, t) = P(\partial/\partial x, \partial/\partial y)\, u(x, y, t) + F(x, y, t) \tag{2.4.1}$$

but now with homogeneous initial data

$$u(x, y, 0) = 0$$

and boundary conditions

$$Lu(0, y, t) = 0, \quad \|u(\cdot, t)\|_{R_d} < \infty.$$

Again, assuming homogeneous boundary data is no real restriction, because otherwise we only need to make a change of the dependent variables.

We are now interested in an estimate for the solution $u(x, y, t)$ in terms of $F(x, y, t)$. We Laplace transform Equations (2.4.1) with respect to t and Fourier transform them with respect to y and obtain the system of ordinary differential equations

$$\begin{aligned} s\hat{u}(x, \omega, s) &= P(\partial/\partial x, i\omega)\, \hat{u}(x, \omega, s) + \hat{F}(x, \omega, s) \\ L\hat{u}(0, \omega, s) &= 0, \end{aligned} \tag{2.4.2}$$

with $\|\hat{u}(\cdot, \omega, s)\| < \infty$ and $s = i\xi + \eta$, $\eta > \eta_0$ sufficiently large. We give an operational definition of well-posedness in the generalized sense:

Definition 2.4.1 *The problem (2.4.1) is called well posed in the generalized sense, if we can estimate the solution* $\hat{u}(x,\omega,s)$ *of the transformed equation (2.4.2) by*

$$\|\hat{u}(\cdot,\omega,s)\| \leq K(\eta)\left\|\hat{F}(\cdot,\omega,s)\right\|, \quad \textit{for } \eta > \eta_0, \tag{2.4.3}$$

where $\|\cdot\|$ *denotes the* L_2*-norm over* x*. For every fixed* η*,* $K(\eta)$ *is a constant which does not depend on* F *and*

$$\lim_{\eta \to \infty} K(\eta) = 0.$$

The estimate (2.4.3) is equivalent with

$$\int_0^\infty e^{-2\eta t}\|u\|_{R_d}^2 dt \leq K^2(\eta)\int_0^\infty e^{-2\eta t}\|F\|_{R_d}^2 dt, \ \eta > \eta_0, \tag{2.4.4}$$

where $\|\cdot\|_{R_d}$ denotes the L_2-norm over all space variables. This follows by applying Parseval's relation (1.1.5):

$$\int_0^\infty e^{-2\eta t}\|u\|_{R_d}^2 dt = \int_0^\infty \|e^{-\eta t}u\|_{R_d}^2 dt = \int_{-\infty}^\infty\int_{-\infty}^\infty \|\hat{u}(x,\omega,i\xi+\eta)\|_x^2 d\omega d\xi.$$

Theorem 2.4.1 *The definition of well-posedness in the generalized sense is stable against lower order perturbations, i.e. if we assume that the problem (2.4.1) is well posed in the generalized sense, then the perturbed problem*

$$v_t(x,y,t) = P(\partial/\partial x, \partial/\partial y, t)\, v(x,y,t) + Bv(x,y,t) + F(x,y,t)$$

is also well posed in the generalized sense, provided B *is a bounded operator.*

Proof. We think of $Bv + F$ as the forcing function. By (2.4.4)

$$\int_0^\infty e^{-2\eta t}\|v\|_{R_d}^2 dt \leq K^2(\eta)\int_0^\infty e^{-2\eta t}\|Bv+F\|_{R_d}^2 dt$$
$$\leq \underbrace{2K^2(\eta)\int_0^\infty e^{-2\eta t}\|Bv\|_{R_d}^2 dt}_{(\star)} + 2K^2(\eta)\int_0^\infty e^{-2\eta t}\|F\|_{R_d}^2 dt.$$

Since $K^2(\eta)$ can be made arbitrary small by choosing η_0 sufficiently large, we can make the term $(\star)$ smaller than half of the left side and get the desired estimate.

Theorem 2.4.2 *If the problem (2.4.1) is well posed, then it is well posed in the generalized sense.*

Proof. For simplicity, we assume that the operator is semibounded, i.e. (2.3.4) holds. We introduce a new variable

$$w(x,y,t) = e^{-\eta t} u(x,y,t),$$

and obtain

$$\frac{\partial}{\partial t} w(x,y,t) = (P - \eta I) w(x,y,t) + e^{-\eta t} F(x,y,t).$$

First, we consider

$$\begin{aligned} \frac{d}{dt}\|w\|^2 &\le (w, w_t) + (w_t, w) \\ &= (w, (P-\eta I) w + e^{-\eta t} F) + ((P-\eta I) w + e^{-\eta t} F, w) \\ &\le -2(\eta - \alpha)\|w\|^2 + 2|(w, e^{-\eta t} F)|. \end{aligned}$$

Since with $\tilde{\eta} = \eta - \alpha$, we get

$$2\left|\left(w, e^{-\eta t} F\right)\right| \le \tilde{\eta}\|w\|^2 + \frac{1}{\tilde{\eta}}\|e^{-\eta t} F\|^2,$$

it follows that

$$\frac{d}{dt}\|w\|^2 \le -\tilde{\eta}\|w\|^2 + \frac{1}{\tilde{\eta}}\|e^{-\eta t} F\|^2.$$

Thus we obtain the following estimate for $\|w\|$ by Gronwall's lemma

$$\|w\|^2 \le \int_0^t e^{-\tilde{\eta}(t-\tau)} \frac{1}{\tilde{\eta}} \|e^{-\eta\tau} F(\cdot,\cdot,\tau)\|^2 d\tau.$$

We define $\varphi(t)$ by

$$\varphi(t) = \begin{cases} e^{-\tilde{\eta} t}, & \text{for } t \ge 0, \\ 0, & \text{for } t < 0. \end{cases}$$

Then, we obtain

$$\begin{aligned} \int_0^\infty e^{-2\eta t}\|u(\cdot,\cdot,t)\|^2 dt &= \int_0^\infty \|w(\cdot,\cdot,t)\|^2 dt \\ &\le \int_0^\infty \left(\int_0^\infty \varphi(t-\tau)\, dt\right) \frac{1}{\tilde{\eta}}\|e^{-\eta\tau} F(\cdot,\cdot,\tau)\| d\tau \\ &\le \frac{1}{\tilde{\eta}^2} \int_0^\infty e^{-2\eta t}\|F(\cdot,\cdot,t)\|^2 dt. \end{aligned}$$ □

Next consider the first order, homogeneous system

$$\begin{aligned} u_t(x,y,t) &= A u_x(x,y,t) + B u_y(x,y,t), \\ u(x,y,0) &= 0, \end{aligned} \tag{2.4.5}$$

with inhomogeneous boundary conditions

$$\begin{gathered} Lu\left(0,y,t\right) =: D^{I}u^{I}\left(0,y,t\right) + D^{II}u^{II}\left(0,y,t\right) = g\left(y,t\right), \\ \left\| u\left(\cdot,t\right)\right\|_{R_d} < \infty. \end{gathered} \tag{2.4.6}$$

Again we Laplace transform with respect to t and Fourier transform with respect to y and obtain

$$\begin{aligned} s\hat{u}\left(x,\omega,s\right) &= A\hat{u}_x\left(x,\omega,s\right) + i\omega B\hat{u}\left(x,\omega,s\right), \\ L\hat{u}\left(0,\omega,s\right) &= \hat{g}\left(\omega,s\right), \end{aligned} \tag{2.4.7}$$

with $\|\hat{u}\| < \infty$.

Definition 2.4.2 *The problem (2.4.5) is called boundary stable if there are constants K, η_0 such that for all g and all real ω and all complex s with* $\operatorname{Re} s \geq \eta_0$

$$|\hat{u}\left(0,s,\omega\right)| < K|\hat{g}\left(s,\omega\right)|.$$

Example. We consider our example (2.2.4) with boundary conditions

$$u\left(0,y,t\right) - \alpha v\left(0,y,t\right) = g\left(y,t\right).$$

Corresponding to (2.2.7), Fourier and Laplace transform give us

$$\begin{gathered} \underline{\hat{u}}_x = M\underline{\hat{u}}, \\ \hat{u}\left(0,\omega,s\right) - \alpha\hat{v}\left(0,\omega,s\right) = \hat{g}\left(\omega,s\right). \end{gathered} \tag{2.4.8}$$

The general solution of (2.4.8) belonging to L_2 is

$$\underline{\hat{u}} = \sigma e^{\kappa_1 x}\begin{pmatrix} s' + \sqrt{s'^2 + \omega'^2} \\ i\omega' \end{pmatrix},$$

with

$$\kappa_1 = -\sqrt{s^2+\omega^2}, \; s' = \frac{s}{|\omega|}, \; \omega' = \frac{\omega}{|\omega|},$$

and σ is determined by the boundary conditions

$$\sigma\left(\left(s' + \sqrt{s'^2+\omega'^2}\right) - i\alpha\omega'\right) = \hat{g}\left(\omega,s\right).$$

Thus the problem is boundary stable, if and only if

$$\min_{s',\omega'}\left|\left(s' + \sqrt{s'^2+\omega'^2}\right) - i\alpha\omega'\right| > 0.$$

Figure 2.3 shows clearly that the problem is boundary stable, if and only if $|\alpha| < 1$.

We rewrite the transformed problem (2.4.7) as follows:

$$\hat{u}_x(x, \omega, s) = M\hat{u}(x, \omega, s),$$

where

$$M = A^{-1}(sI - i\omega B).$$

It follows from Schur's Lemma (1.4.2) that there exists a unitary matrix U, $U^*U = I$ such that

$$U^*MU = \begin{pmatrix} M_{11} & M_{12} \\ 0 & M_{22} \end{pmatrix}$$

is in upper triangular form and the eigenvalues κ of M_{11} have negative real parts and the eigenvalues κ of M_{22} have positive real parts. By the change of variables $\hat{u} = U\hat{v}$ we obtain from (2.4.5)

$$\hat{v}_x(x, \omega, s) = \begin{pmatrix} M_{11} & M_{12} \\ 0 & M_{22} \end{pmatrix} \hat{v}(x, \omega, s).$$

The boundary conditions become

$$\tilde{D}^I \hat{v}^I(0, \omega, s) + \tilde{D}^{II} \hat{v}^{II}(0, \omega, s) = \hat{g}(\omega, s).$$

Since we required that the solution is in L_2, it follows that $\hat{v}^{II}(0, \omega, s) \equiv 0$, because the eigenvalues of M_{22} have positive real parts. So we have

$$\hat{v}^I(x, \omega, s) = e^{M_{11}x}\hat{v}(0, \omega, s)$$

and

$$\tilde{D}^I \hat{v}^I(0, \omega, s) = \hat{g}(\omega, s).$$

Therefore the assumption that the problem is boundary stable is equivalent with the request $\left|\left(\tilde{D}^I\right)^{-1}\right| \leq K$. We observe that our test in Section 2.2 requires that $\det(\tilde{D}^I)^{-1} \neq 0$ but no estimate is required.

Theorem 2.4.3 *Assume that the system (2.4.1) is symmetric hyperbolic or strongly hyperbolic and the eigenvalues of the symbol $P(i\omega) = iA\omega_1 + iB\omega_2$ have constant multiplicity.*

If the problem is boundary stable, then it is well posed in the generalized sense.

Remark. The assumption that a problem is boundary stable is slightly more restrictive than well-posedness. In Example 1, we saw that the problem

$$\underline{u}_t = \begin{pmatrix} -1 & 0 \\ 0 & +1 \end{pmatrix} \underline{u}_x + \begin{pmatrix} 0 & 1 \\ 1 & 0 \end{pmatrix} \underline{u}_y$$

with boundary conditions

$$u^{(1)}(0, y, t) = \alpha u^{(2)}(0, y, t)$$

is not well posed, if $|\alpha| > 1$ and $Im(\alpha) \neq 0$. One can show that it is well posed in the generalized sense, if $|\alpha| \leq 1$, and that it is not well posed in the generalized sense, if α is real and $|\alpha| > 1$. As we have shown earlier, the problem is boundary stable, if $|\alpha| < 1$.

Proof. We prove the result only for symmetric systems, i.e. we assume that $A = A^*$ and $B = B^*$. We consider the inhomogeneous system (2.4.5)

$$u_t(x, y, t) = Au_x(x, y, t) + Bu_y(x, y, t) + F(x, y, t), \tag{2.4.9}$$

with homogeneous boundary conditions

$$\begin{gathered} Lu(0, y, t) =: D^I u^I(0, y, t) + D^{II} u^{II}(0, y, t) = 0, \\ \|u(\cdot, t)\|_{R_d} < \infty. \end{gathered} \tag{2.4.10}$$

As in (2.2.1), A is nonsingular and diagonal.

By Laplace transform and Fourier transform of (2.4.9) and (2.4.10), we obtain

$$\begin{aligned} s\hat{u}(x, \omega, s) &= A\hat{u}_x(x, \omega, s) + i\omega B\hat{u}(x, \omega, s) + \hat{F}(x, \omega, s), \\ L\hat{u}(0, \omega, s) &= 0. \end{aligned}$$

First we solve the auxiliary problem

$$\begin{aligned} s\hat{v}(x, \omega, s) &= A\hat{v}_x(x, \omega, s) + i\omega B\hat{v}(x, \omega, s) + \hat{F}(x, \omega, s), \\ \hat{v}^I(0, \omega, s) &= 0. \end{aligned}$$

Multiplying by $\hat{v}$ and taking the L_2-norm over x leads to

$$\operatorname{Re} s(\hat{v}, \hat{v}) = \operatorname{Re}(\hat{v}, A\hat{v}_x) + \operatorname{Re}(\hat{v}, i\omega B\hat{v}) + \operatorname{Re}\left(\hat{v}, \hat{F}\right). \tag{2.4.11}$$

Since B is symmetric $\operatorname{Re}(\hat{v}, i\omega B\hat{v}) = 0$. Integration by parts yields

$$(\hat{v}, A\hat{v}_x) = -\left.\langle \hat{v}, A\hat{v}\rangle\right|_{x=0} - (\hat{v}_x, A\hat{v}),$$

and since A is symmetric and $\hat{v}^I(0,\omega,s)=0$

$$2\ \mathrm{Re}\,(\hat{v},A\hat{v}_x) = -\left.\left\langle \hat{v}^{II},\Lambda_2\hat{v}^{II}\right\rangle\right|_{x=0}.$$

Thus we obtain from (2.4.11) by using the Cauchy-Schwarz inequality

$$\mathrm{Re}\,s\|\hat{v}\|^2 + \frac{1}{2}\left.\left\langle \hat{v}^{II},\Lambda_2\hat{v}^{II}\right\rangle\right|_{x=0} \le \|\hat{v}\|\|\hat{F}\|. \tag{2.4.12}$$

Since the entries of Λ_2 are positive, it follows that there is a constant $\delta>0$ such that

$$\left.\left\langle \hat{v}^{II},\Lambda_2\hat{v}^{II}\right\rangle\right|_{x=0} \ge \delta^2 \left.\left|\hat{v}^{II}\right|^2\right|_{x=0}.$$

Thus we obtain from (2.4.12) the estimates

$$\|\hat{v}\| \le \frac{1}{\mathrm{Re}\,s}\|\hat{F}\|, \tag{2.4.13}$$

and at the boundary

$$\delta|\hat{v}|^2 = \delta|\hat{v}^{II}|^2 \le \left\langle \hat{v}^{II},\Lambda_2\hat{v}^{II}\right\rangle^{1/2} \le \frac{1}{\mathrm{Re}\,s}\|\hat{F}\|. \tag{2.4.14}$$

Now we consider $\hat{w}=\hat{u}-\hat{v}$, which satisfies

$$\begin{aligned} s\hat{w}(x,\omega,s) &= A\hat{w}_x(x,\omega,s) + i\omega B\hat{w}(x,\omega,s), \\ L\hat{w}(0,\omega,s) &= -L\hat{v}(0,\omega,s). \end{aligned} \tag{2.4.15}$$

By (2.4.14), we have for $\hat{v}$ the estimate

$$|L\hat{v}(x,\omega,s)| \le const\frac{\|\hat{F}\|}{\delta\,\mathrm{Re}\,s}.$$

Since by assumption, the problem (2.4.15) is boundary stable, we obtain

$$|\hat{w}(0,\omega,s)| \le const\frac{K\|\hat{F}\|}{\mathrm{Re}\,s}.$$

Thus we have an estimate of $\hat{w}$ at the boundary. In the same way as before, we can estimate $\|\hat{w}(\cdot,\omega,s)\|$. Corresponding to (2.4.11), we obtain

$$\mathrm{Re}\,s\,(\hat{w},\hat{w}) = \mathrm{Re}\,(\hat{w},A\hat{w}_x) + \mathrm{Re}\,(\hat{w},i\omega\hat{w}).$$

Integration by parts gives us

$$\begin{aligned} \mathrm{Re}\,s\,\|\hat{w}(\cdot,w,s)\|^2 &= -\frac{1}{2}\left.\langle \hat{w}(\cdot,w,s), A\hat{w}(\cdot,w,s)\rangle\right|_{x=0} \\ &\le \frac{const\,K^2}{\delta^2(\mathrm{Re}\,s)^2}\left\|\hat{F}(\cdot,w,s)\right\|^2. \end{aligned}$$

Therefore by (2.4.13)

$$\|\hat{u}(\cdot, w, s)\| \leq \|\hat{v}(\cdot, w, s)\| + \|\hat{w}(\cdot, w, s)\| \leq \frac{const}{\operatorname{Re} s + (\operatorname{Re} s)^{3/2}} \|\hat{F}(\cdot, w, s)\|,$$

and we have proved that the problem is well posed in the generalized sense.

Remark. If A and B are not symmetric, but the system is strongly hyperbolic and the eigenvalues of the symbol $P(i\omega)$ have constant multiplicity, the proof follows the same arguments but in another norm.

2.5 Farfield Boundary Conditions

Consider again the problem

$$u_t(x, y, t) = Au_x(x, y, t) + Bu_y(x, y, t)$$

in the half plane $x \geq 0$, $y \in \mathbb{R}$. We again assume that A is a diagonal matrix

$$A = \begin{pmatrix} -\Lambda_1 & 0 \\ 0 & \Lambda_2 \end{pmatrix}, \quad \Lambda_1 > 0,\ \Lambda_2 > 0.$$

The boundary conditions at $x = 0$ are given by

$$u^I(0, y, t) = u_0^I(y, t),$$

where $u = \left(u^I, u^{II}\right)^T$. We can solve the problem and obtain u^{II} at the boundary as a function of u_0^I

$$u^{II}(0, y, t) = Ku^I(0, y, t) = Ku_0^I(y, t).$$

Here K is a convolution operator which we can determine by Laplace and Fourier transform. We shall explain the procedure by using our previous example.

$$u_t(x, y, t) = \begin{pmatrix} -1 & 0 \\ 0 & 1 \end{pmatrix} u_x(x, y, t) + \begin{pmatrix} 0 & 1 \\ 1 & 0 \end{pmatrix} u_y(x, y, t),$$

with homogeneous initial data $u(x, y, 0) = 0$ and boundary conditions

$$u^I(0, y, t) = u_0^I(y, t).$$

As before, we obtain by Laplace and Fourier transform

$$s\hat{u}(x, \omega, s) = \begin{pmatrix} -1 & 0 \\ 0 & 1 \end{pmatrix} \hat{u}_x(x, \omega, s) + i\omega B\hat{u}(x, \omega, s)$$

with boundary conditions

$$\hat{u}^I(0,\omega,s) = \hat{u}_0^I(\omega,s),$$

which is solved by

$$\hat{u}(x,\omega,s) = \sigma e^{-\sqrt{s^2+\omega^2}\,x}\begin{pmatrix} s+\sqrt{s^2+\omega^2} \\ i\omega \end{pmatrix}.$$

The parameter σ is determined by the boundary conditions:

$$\sigma\left(s+\sqrt{s^2+\omega^2}\right) = \hat{u}_0^I(\omega,s).$$

Thus we obtain the relation

$$\hat{u}^{II}(0,\omega,s) = i\omega\sigma = \frac{i\omega}{s+\sqrt{s^2+\omega^2}}\hat{u}_0^I(\omega,s). \tag{2.5.1}$$

In physical space, the relation (2.5.1) is an integral relation and thus non-local. It is very complicated to effectively compute $\hat{v}(0,\omega,s)$ in terms of $\hat{u}_0(\omega,s)$. Therefore, one uses asymptotic expansions to simplify the computation. This is only possible, if one restricts the data. We consider boundary data with

$$\begin{aligned} &\hat{u}_0^I(\omega,s) \neq 0 \quad \text{for } |s| \gg |\omega|, \\ &\hat{u}_0^I(\omega,s) = 0 \quad \text{otherwise.} \end{aligned}$$

Since

$$s+\sqrt{s^2+\omega^2} = s\left(1+\sqrt{1+\left(\frac{\omega}{s}\right)^2}\right) = s\left(2+\frac{1}{2}\frac{\omega^2}{s^2}+\ldots\right),$$

we can write (2.5.1) as

$$\hat{u}^{II}(0,\omega,s) = \frac{i\omega}{2s+\frac{1}{2}\frac{\omega^2}{s}+\ldots}\hat{u}^I(0,\omega,s).$$

To first approximation, the last relation can be replaced by

$$\hat{u}^{II}(0,\omega,s) = 0,$$

or in physical space by

$$u^{II}(0,y,t) = 0.$$

A better approximation is given by

$$2s\hat{u}^{II}(0,\omega,s) = i\omega\hat{u}^I(0,y,t).$$

Since s corresponds to $\partial/\partial t$ and $i\omega$ to $\partial/\partial y$, the last relation becomes

$$2u_t^{II}(0,y,t) = u_y^I(0,y,t).$$

Similar, one can make an expansion using the assumption $|\omega| \ll |s|$, where the problem is almost stationary. This leads to different relations in the boundary.

Next we consider a system with variable coefficients

$$u_t(x,y,t) = A(x,y,t)\,u_x(x,y,t) + B(x,y,t)\,u_y(x,y,t), \tag{2.5.2}$$

for $x \ge 0$ and $0 < y < 1$. We assume that A and B are 1-periodic in y and converge for $x \to \infty$ to constant matrices A_0, B_0. Typically

$$\begin{aligned} A(x,y,t) &= A_0 + \frac{1}{(x+1)^p} A_1(x,y,t), \quad p \ge 1, \\ B(x,y,t) &= B_0 + \frac{1}{(x+1)^p} B_1(x,y,t). \end{aligned}$$

We are interested in solutions which are 1-periodic in y and satisfy boundary conditions

$$Lu(0,y,t) = g(y,t), \quad \|u(\cdot,t)\| < \infty.$$

Since the coefficients are variable, we cannot in general calculate the solution analytically. However, in a numerical calculation, one can only treat problems in bounded domains. Therefore we proceed in the following way.

We divide the region into two regions R_1, R_2, defined by

$$\begin{aligned} R_1 &: \ (0 \le x \le l, 0 \le y \le 1), \\ R_2 &: \ (l \le x \le \infty, 0 \le y \le 1). \end{aligned}$$

We choose $l \gg 1$ (see Figure 2.5). Since A, B converge to A_0, B_0, we neglect A_1, B_1 and solve in the outer region the system

$$u_t^{(0)} = A_0 u_x^{(0)} + B_0 u_y^{(0)}. \tag{2.5.3}$$

Assuming that

$$A_0 = \begin{pmatrix} -\Lambda_1 & 0 \\ 0 & \Lambda_2 \end{pmatrix}, \quad \Lambda_1 > 0, \ \Lambda_2 > 0,$$

is a diagonal matrix, we solve (2.5.3) with boundary conditions

$$\left(u^{(0)}\right)^I (l,y,t) = g^{(0)}(y,t),$$

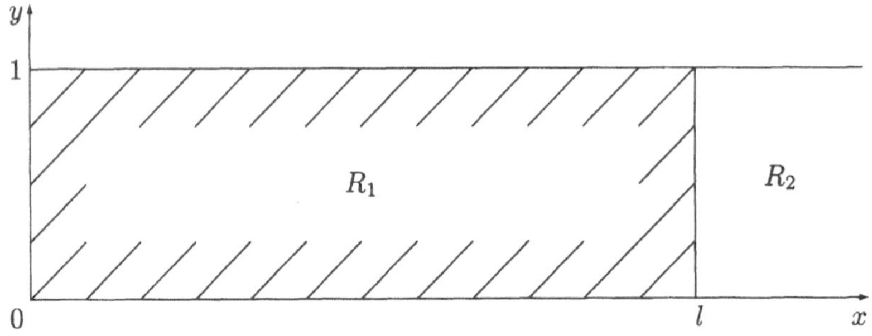

Figure 2.5: Near- and far-field for example 2.5.2

where $g^{(0)}$ is the constant term in the expansion of g. By our previous arguments, we can determine $\left(u^{(0)}\right)^{II}(l, y, t)$. We can easily derive relations between

$$\left(\hat{u}^{(0)}\right)^{I}(l, \omega, s) \text{ and } \left(\hat{u}^{(0)}\right)^{II}(l, \omega, s)$$

on the transform side. (Instead of the Fourier transform in y we use Fourier series.)

Using asymptotic expansions, we can translate these relations approximately into differential relations in physical space. We use these relations as boundary conditions for our calculations in the bounded domain R_1.

The above problem has been treated extensively. We refer to a survey article by T. Hagström [2] and work by M. Grote and J. Keller [4].

2.6 Energy Estimates

We considered different techniques to investigate boundary conditions: For one-dimensional problems $u_t = Au_x$, the methods of characteristics leads to $u^I = Su^{II}$. For two-dimensional problems $u_t = Au_x + Bu_y$, the number of boundary conditions is determined by the number of negative characteristics. But not all relationes of the form $u^I = Su^{II}$ lead to well-posed problems. In this case we try to find bad test solutions by using Fourier and Laplace transform.

Next we introduce an intermediate technique, the energy estimates. This technique does not always work, but if it works, it is quite simple and effective and thus should be checked first.

For the problem

$$u_t(x,t) = P(\partial/\partial x)\, u(x,t), \tag{2.6.1}$$

we want an estimate of the form

$$\frac{\partial}{\partial t}\|u\|^2 \le \alpha \|u\|^2. \tag{2.6.2}$$

We consider the problem $u_t(x,t) = Au_x(x,t)$ and assume that $A = A^*$:

$$\begin{aligned}
\frac{\partial}{\partial t}(u,u) &= (u,u_t) + (u_t,u) \\
&= (u,Au_x) + (Au_x,u) \\
&= -(Au_x,u) - \langle u, Au\rangle|_{x=0} + (Au_x,u) \\
&\le 0,
\end{aligned}$$

if $Lu(0,t) = 0$. In this case, the estimate (2.6.2) is valid with $\alpha = 0$. If this technique does not work, one might be able to define a new norm for which it works.

2.7 First Order Systems with Variable Coefficients

We consider a first order system

$$\begin{aligned}
u_t(x,y,t) &= A(x,y,t)\, u_x(x,y,t) + B(x,y,t)\, u_y(x,y,t), \\
u(x,y,0) &= f(x,y),
\end{aligned} \tag{2.7.1}$$

with smooth coefficients in some domain $\Omega \times [0,\infty[$ with smooth boundary $\partial\Omega$. At every boundary point we give boundary conditions

$$L(x,y,t)\, u = 0 \tag{2.7.2}$$

consisting of smooth linear relations between the components of u. We assume that the Cauchy problem for (2.7.1) is well posed. At every boundary point we freeze the coefficients of the differential equation and the boundary conditions and rotate the coordinate system such that the inner normal points in the direction of the positive x-axis. We can now define for every boundary point a halfplane problem with constant coefficients. If all these halfplane problems are well posed in the generalized sense and satisfy the conditions of this section, then the same is true for the linear problem (2.7.1), (2.7.2) with variable coefficients.

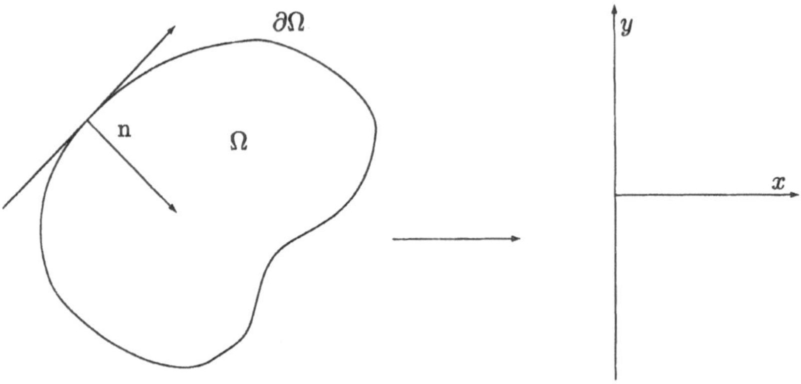

Figure 2.6: We consider every boundary point.

2.8 Remarks

Detailed discussions of the material of Chapter 2 are contained in Chapter 7 of Kreiss and Lorenz and Chapters 9 and 10 of Gustafsson, Kreiss and Oliger.

Chapter 3

Difference Methods

3.1 Periodic Problems

Consider the problem

$$u_t(x,t) = u_x(x,t), \quad -\infty \le x < \infty,\ t \ge 0, \tag{3.1.1}$$

with initial data

$$u(x,0) = f(x). \tag{3.1.2}$$

We assume that $f(x)$ and therefore also the solution of (3.1.1) is 2π-periodic. We discretize Equation (3.1.1) in space but keep time continuous.

We denote by

$$h = \frac{2\pi}{2N+1}, \quad N \in \mathbb{N},$$

a grid interval and define the corresponding uniform grid by

$$x_\nu = \nu h, \quad \nu = 0, \pm 1, \pm 2, \ldots$$

A discrete function

$$v_\nu = v(x_\nu)$$

defined on the grid is called gridfunction. The translation operator E is defined for $p = 0, \pm 1, \pm 2, \ldots$ by

$$E^p v_\nu = v_{\nu+p}.$$

Central, forward and backward difference operators are defined by

$$D_0 = \frac{1}{2h}(E - E^{-1})$$

and

$$D_+ = \frac{1}{h}(E-I), \quad D_- = \frac{1}{h}(I-E).$$

The perhaps simplest difference approximations of (3.1.1) are

$$v_{\nu t} = D_- v_\nu, \tag{3.1.3}$$

$$v_{\nu t} = D_+ v_\nu, \tag{3.1.4}$$

$$v_{\nu t} = D_0 v_\nu. \tag{3.1.5}$$

The first two approximations are first and the third approximation is second order accurate respectively. One can of course also construct higher order methods. For example

$$v_{\nu t} = \left(D_0 - \frac{h^2}{6} D_0 D_+ D_- \right) v_\nu$$

is a fourth order accurate method. The general approximation of (3.1.1) is of the form

$$v_{\nu t} = Q\,(E)\, v_\nu, \tag{3.1.6}$$

where Q is a polynomial in E and E^{-1}. We now proceed in exactly the same way as for the differential equations. We look for simple wave solutions

$$v_\nu\,(t) = e^{i\omega x_\nu} \hat{v}\,(t). \tag{3.1.7}$$

Since

$$D_-\, e^{i\omega x_\nu} = e^{i\omega x_\nu} \left(\frac{1-e^{-i\omega h}}{h} \right) = e^{i\omega x_\nu} \left(\frac{1-\cos\omega x}{h} + i\,\frac{\sin\omega x}{h} \right),$$

$$D_+\, e^{i\omega x_\nu} = e^{i\omega x_\nu} \left(\frac{e^{-i\omega h}-1}{h} \right) = e^{i\omega x_\nu} \left(\frac{\cos\omega x - 1}{h} + i\,\frac{\sin\omega x}{h} \right),$$

$$D_0\, e^{i\omega x_\nu} = e^{i\omega x_\nu} \left(\frac{e^{i\omega h} - e^{-i\omega h}}{2h} \right) = i\,\frac{\sin\omega h}{h} e^{i\omega x_\nu},$$

we obtain for the approximations (3.1.3)–(3.1.5)

$$\hat{v}_t = \left(\frac{1-\cos\omega h}{h} + i\,\frac{\sin\omega h}{h} \right) \hat{v}, \tag{3.1.8}$$

$$\hat{v}_t = \left(\frac{\cos\omega h - 1}{h} + i\,\frac{\sin\omega h}{h} \right) \hat{v}, \tag{3.1.9}$$

$$\hat{v}_t = i\,\frac{\sin\omega h}{h} \hat{v}. \tag{3.1.10}$$

(3.1.8)–(3.1.10) are ordinary differential equations. Since

$$\frac{1-\cos \omega h}{h}=\frac{1}{h} \quad \text{for} \quad \omega=\frac{\pi}{2h},$$

there are simple wave solutions of (3.1.8) which grow arbitrarily fast for $h \to 0$. Therefore the approximation is useless and called unstable. The solutions of (3.1.9), (3.1.10) do not grow and are called stable. Applying our test to the general approximation (3.1.6) gives us

$$\hat{v}_t(t)=\hat{Q}\left(e^{i\omega h}\right)\hat{v}(t),$$

i.e.

$$\hat{v}(t)=e^{\hat{Q}\left(e^{i\omega h}\right)t}\hat{v}(0).$$

The principle of well-posedness is now replaced by stability.

Definition 3.1.1 *We call the approximation stable, if*

$$|\hat{v}(t)| \le Ke^{\alpha t}|\hat{v}(0)|, \quad \text{i.e.} \quad \left|e^{\hat{Q}\left(e^{i\omega h}\right)t}\right| \le Ke^{\alpha t}$$

for h sufficiently small ($h \le h_0$) and K and α independent of ω and h.

We observe that the approximations (3.1.4), (3.1.5) are stable with $\alpha = 0$ and $K = 1$. Thus stability is equivalent with an estimate of the form

$$\left|e^{\hat{Q}\left(e^{i\omega h}\right)t}\right| \le Ke^{\alpha t}.$$

Besides stability the concept of dissipation is important. There are mainly two reasons for introducing dissipative schemes.

1. For $|\omega h| > \delta$, we have

   $$\frac{i\sin(\omega h)}{h} \not\approx i\omega$$

   and thus the above approximations become useless for these frequencies. Dissipation can kill the high frequency terms.
2. When solutions with discontinuities arise, dissipation can suppress undesirable oscillations.

The simplest way to obtain a dissipative scheme is to add a second order difference term. For example we replace (3.1.5) by

$$v_{\nu t}=D_0 v_\nu+\sigma h D_+ D_- v_\nu,$$

where σ is a positive constant. The transformed equation becomes

$$\hat{v}_t(t) = \left(\frac{i\sin\omega h}{h} - 4\frac{\sigma}{h}\sin^2\frac{\omega h}{2}\right)\hat{v}(t) \tag{3.1.11}$$

with initial data

$$\hat{v}_\nu(0) = \hat{f}_\nu.$$

For $\delta \leq |\omega h| \leq 2\pi - \delta$ the solutions of (3.1.11) decay rapidly. For $|\omega h| \ll 1$ Taylor expansion gives us

$$\hat{v}_t = (i\omega - \sigma h\omega^2 + O(h^2\omega^3))\hat{v},$$

which converges linearly in h to $\hat{v}_t = i\omega\hat{v}$.

A more accurate approximation is

$$v_{\nu t} = D_0 v_\nu - \sigma h^3 D_+^2 D_-^2 v_\nu.$$

Now the transformed equation becomes

$$\hat{v}_t(t) = \left(\frac{i\sin\omega h}{h} - 16\frac{\sigma}{h}\sin^4\frac{\omega h}{2}\right)\hat{v}(t).$$

For $\delta \leq |\omega h| \leq 2\pi - \delta$ its solutions decay rapidly, while for $|\omega h| \ll 1$ Taylor expansion gives us

$$\hat{v}_t = \left(i\omega - \frac{1}{6}\omega^3 h^2 - \sigma\omega^4 h^3 + O(\omega^5 h^4)\right)\hat{v},$$

which converges quadratically in h to $\hat{v}_t = i\omega\hat{v}$. Since $\sin^2\frac{\omega h}{2} = 0$ for $\omega h = 2\pi$ there is no effect of the dissipation for this frequency. However, we interpolate the initial data and therefore also the solution by a Fourier polynomial

$$f_\nu = \sum_{\omega=-N}^{N} \tilde{f}(\omega)\, e^{i\omega x_\nu},$$

where $|\omega h| \leq Nh \leq \frac{2\pi N}{2N+1} < \pi$. Thus we have no problem with the dissipation vanishing for high frequencies. We formalize our discussion in

Definition 3.1.2 *Consider the general approximation (3.1.7) of (3.1.1). It is dissipative of order p if*

$$\operatorname{Re}\hat{Q}\left(e^{i\omega h}\right) \leq \frac{-\delta|\omega h|^p}{h}$$

for $|\omega| \leq \pi$.

There are no problems to generalize our results to systems in many space dimensions. As an example we consider a strictly hyperbolic first order system with constant coefficients

$$u_t(x,y,t) = Au_x(x,y,t) + Bu_y(x,y,t) \tag{3.1.12}$$

with 2π-periodic initial data

$$u(x,y,0) = f(x,y). \tag{3.1.13}$$

The grid points are now defined by

$$x_\nu = \nu h,\ y_\mu = \mu h,\ \nu = 0, \pm 1, \dots,\ \mu = 0, \pm 1, \dots$$

Denoting by D_{+x}, D_{-x}, D_{0x}, D_{+y}, D_{-y}, D_{0y}, the difference operators in the x and y direction respectively we approximate (3.1.12), (3.1.13) by

$$\begin{aligned} v_t(x_\nu, y_\mu, t) &= \left((AD_{0x} + BD_{0y}) - \sigma h^3 \left(D_{+x}^2 D_{-x}^2 + D_{+y}^2 D_{-y}^2\right)\right) v(x_\nu, y_\mu, t), \\ v(x_\nu, y_\mu, 0) &= f(x_\nu, y_\mu, 0). \end{aligned} \tag{3.1.14}$$

Introducing the test function

$$v = e^{i(\omega_1 x + \omega_2 y)} \hat{v}(\omega, t)$$

gives us

$$\hat{v}_t = \hat{Q}(\omega h) \hat{v},$$

where

$$\hat{Q}(\omega h) = \frac{i}{h}(A \sin \omega_1 h + B \sin \omega_2 h) - 16\frac{\sigma}{h}\left(\sin^4 \frac{\omega_1 h}{2} + \sin^4 \frac{\omega_2 h}{2}\right) I.$$

Since by assumption the system is strictly hyperbolic we can find for every fixed $h\omega = (h\omega_1, h\omega_2)$ a transformation S such that

$$S^{-1}\hat{Q}S = \frac{i}{h}\Lambda - \frac{16\sigma}{h}\left(\sin^4 \frac{\omega_1 h}{2} + \sin^4 \frac{\omega_2 h}{2}\right) I.$$

Here Λ is a real diagonal matrix. Also, there is a constant K such that for all ωh

$$|S^{-1}||S| \le K.$$

We can now prove

Theorem 3.1.1 *The approximation (3.1.14) is stable.*

Proof.

$$\left|e^{\hat{Q}(\omega h)t}\right| = \left|SS^{-1}e^{\hat{Q}(\omega h)t}SS^{-1}\right| \le |S||S^{-1}|\left|e^{S^{-1}\hat{Q}(\omega h)St}\right| \le |S||S^{-1}|. \quad \square$$

One can of course construct many different difference approximations of (3.1.12). Typically they are of the form

$$v_t\left(x_\nu, y_\mu, t\right) = \frac{1}{h} Q\left(hD_+, hD_-, hD_0\right) v\left(x_\nu, y_\mu, t\right), \tag{3.1.15}$$

where Q is a polynomial in hD_+, hD_- and hD_0. The transformed system becomes

$$\hat{v}_t = \frac{1}{h}\hat{Q}\left(\omega h\right)\hat{v}, \tag{3.1.16}$$

where $\hat{Q}$ is a uniformly bounded smooth matrix function of $h\omega$. The generalization of Definition 3.1.2 is

Definition 3.1.3 *The approximation (3.1.15) is dissipative of order p, if the eigenvalues $\lambda_k\left(\omega h\right)$ of $\hat{Q}\left(\omega h\right)$ satisfy the inequality*

$$\operatorname{Re}\lambda_k\left(\omega h\right) \le -\frac{\delta}{h}|h\omega|^p, \ |h\omega| \le \pi, \ \delta = const. > 0.$$

Clearly the approximation (3.1.14) is dissipative of order 4.

The main result of this section is

Theorem 3.1.2 (Stability for hyperbolic problems) *Consider a strictly hyperbolic first order system of partial differential equations. If the difference approximation is consistent (i.e. if the solution of the difference problem converges formally to the solution of the differential equation) and dissipative of any order $p > 0$, then the approximation is stable.*

Proof. Let τ be any fixed constant. For $|\omega h| \ge \tau$, $\operatorname{Re}\lambda_k\left(\omega h\right)$ is strictly negative. Therefore there is a constant K_1 such that

$$\sup_{|\omega h| \ge \tau, t}\left|e^{\frac{1}{h}\hat{Q}(\omega h)t}\right| \le K_1.$$

Consistency implies that (3.1.15) is of the form

$$v_t = AD_{0x}v + BD_{0y}v + hQ_1 v,$$

where $hQ_1 v$ represents higher order terms. (Compare with (3.1.14).) Therefore for $|\omega h| \ll 1$ we can write (3.1.16) in the form

$$\hat{v}_t = |\omega|\left\{i\left(A\omega_1' + B\omega_2'\right) + hC\left(\omega_1', \omega_2'\right)\right\}\hat{v}, \ \omega' = \omega/|\omega|.$$

Here C is some matrix function of ω_1', ω_2'. Since by assumption the eigenvalues of $A\omega_1' + B\omega_2'$ are distinct we can diagonalize the system by a uniformly bounded transformation. The eigenvalues have, by assumption, negative real parts and therefore stability follows. □

The simplest parabolic equation is the heat equation

$$u_t(x,t) = u_{xx}(x,t) \tag{3.1.17}$$

Its Fourier transform is given by

$$\hat{u}_t(\omega,t) = -\omega^2 \hat{u}(\omega,t).$$

In contrast to hyperbolic, parabolic equations are "naturally" dissipative. They represent a smoothing process. We approximate (3.1.17) by

$$v_t(x_\nu,t) = D_+D_-v(x_\nu,t).$$

The transformed equation is

$$\hat{v}_t = -\frac{4}{h^2}\sin^2\frac{\omega h}{2}\hat{v}.$$

Since

$$\frac{4}{h^2}\sin^2\frac{\omega h}{2} \geq \frac{2}{\pi}\omega^2 \quad \text{for } |\omega h| \leq \pi$$

the approximation is also dissipative. This is typical for all used approximations of parabolic systems.

Let Q denote a difference operator and consider the approximation of a second order parabolic system

$$v_t(x,t) = Q(E)v(x,t) \tag{3.1.18}$$

and its Fourier transform

$$\hat{v}_t(\omega,t) = \hat{Q}(\omega h)\hat{v}(\omega,t).$$

Definition 3.1.4 *The approximation (3.1.18) is called parabolic, if the eigenvalues of $\hat{Q}$ satisfy the estimate*

$$\operatorname{Re}\lambda\left(\hat{Q}\right) \leq -\delta|\omega|^2, \quad \delta = cons. > 0$$

for all $|\omega h| \leq \pi$.

In the same way as Theorem 3.1.2 one can prove

Theorem 3.1.3 (Stability for parabolic problems) *If the approximation of a parabolic equation is parabolic, then the approximation is stable.*

In the same way as for ordinary differential equations we can use stability combined with truncation error analysis to derive error estimates. For wave propagation problems the concept of "points per wavelength" often gives a better way to determine the necessary grid length h. We shall discuss the concept with help of our standard model problem. Consider

$$\begin{aligned} u_t(x,t) &= u_x(x,t), \\ u(x,0) &= e^{i\omega x}, \end{aligned} \tag{3.1.19}$$

and approximate it by

$$\begin{aligned} v_t(x_\nu,t) &= D_0 v(x_\nu,t), \\ v(x_\nu,0) &= e^{i\omega x_\nu}. \end{aligned} \tag{3.1.20}$$

The solutions of the above problems are given by

$$u(x,t) = e^{i\omega(t+x)}, \quad v(x_\nu,t) = e^{i\omega(\alpha t+x_\nu)}, \quad \alpha = \frac{\sin\omega h}{\omega h},$$

respectively. Let E denote the difference between the solutions. Then

$$E = |u(x_\nu,t) - v(x_\nu,t)| = |e^{i\omega t} - e^{i\omega\alpha t}| = |1 - e^{i\omega(1-\alpha)t}|.$$

Thus E is approximately

$$E \approx |\omega(1-\alpha)t| = |\omega t||1 - \frac{\sin(\omega h)}{\omega h}|.$$

By Taylor expansion, we obtain

$$E \approx |\omega t|\frac{\omega h}{6}.$$

We want to express E in terms of points per wavelength. We denote by M the number of points in the wave with wave number ω

$$M = \frac{2\pi}{h\omega},$$

and by

$$q = \frac{\omega t}{2\pi}$$

Table 3.1: Number of points per wavelength for a given tolerance E.

	M_2	M_4	M_6
$E = 0.1$	$20q^{1/2}$	$7q^{1/4}$	$5q^{1/6}$
$E = 0.01$	$64q^{1/2}$	$13q^{1/4}$	$8q^{1/5}$

the number of time periods we want to calculate. Then

$$E = \frac{2\pi q}{6} \frac{(2\pi)^2}{M^2}. \tag{3.1.21}$$

With the formula (3.1.21) we can calculate the error E for any given number of points per wavelength. In practice, the other way round is more interesting. We want to know how many points we need such that the error is below a given tolerance. Solving (3.1.21) for M gives us

$$M_2 = \sqrt{\frac{\pi}{3}}\, 2\pi \left(\frac{q}{E}\right)^{1/2}.$$

Instead of approximating (3.1.19) by a second order method we can approximate it by a fourth or higher order method. The usual centered fourth order method is given by

$$v_t\left(x_\nu, t\right) = \left(D_0 - \frac{h^2}{6} D_0 D_+ d_-\right) v\left(x_\nu, t\right).$$

In this case we obtain

$$M_4 = 2\pi \left(\frac{\pi}{14}\right)^{1/4} \left(\frac{q}{E}\right)^{1/4}.$$

For the corresponding 6th order method

$$M_6 = 2\pi \left(\frac{\pi}{70}\right)^{1/6} \left(\frac{q}{E}\right)^{1/4}.$$

In Table 3.1 we have calculated the number of points/wavelength such that the error is below a given tolerance. I shows clearly that fourth order schemes are much superior to second order schemes for wave propagation problems. The improvement from fourth to sixth order schemes is not so pronounced.

3.2 Half Plane Problems

Consider the half plane problem

$$u_t(x,t) = u_x(x,t), \quad x \ge 0,\ t \ge 0, \tag{3.2.1}$$

with initial data

$$u(x,0) = f(x).$$

We assume that $f(x) \in L_2$, i.e.

$$\|f(\cdot)\|^2 =: \int_0^\infty |f(x)|^2 dx < \infty,$$

and are interested in solutions which belong to L_2 for every fixed t. No boundary conditions are necessary since at $x = 0$ the characteristics leave the domain. We can use integration by parts to derive an energy estimate. With the notation

$$(f,g) = \int_0^\infty \bar{f} g dx,$$

we use the identity

$$\begin{aligned}(\partial f/\partial x, g) &= \int_0^\infty \partial \bar{f}/\partial x g dx \\ &= -\int_0^\infty \bar{f} \partial g/\partial x dx - \bar{f} g|_{x=0} \\ &= (f, \partial g/\partial x) - \bar{f}(0) g(0)\end{aligned} \tag{3.2.2}$$

to obtain from (3.2.1)

$$\begin{aligned}\frac{\partial}{\partial t}\|u\|^2 &= (u_t, u) + (u, u_t) \\ &= (u_x, u) + (u, u_x) \\ &= -(u, u_x) + (u, u_x) - |u(0,t)|^2 \\ &= -|u(0,t)|^2 \\ &\le 0,\end{aligned}$$

i.e. for every t

$$\|u(\cdot,t)\|^2 \le \|u(\cdot,0)\|^2. \tag{3.2.3}$$

We approximate (3.2.1) by

$$v_{\nu t}(t) = D_0 v_\nu(t), \quad \nu = 0,1,2,\ldots \tag{3.2.4}$$

In order to solve Equation (3.2.4) on the boundary $x = 0$, we have to introduce a ghost point x_{-1} and add an extrapolation condition.

$$(hD_+)^p \, v_{-1}(t) = 0. \tag{3.2.5}$$

For $p = 2$ this would be

$$h^2 D_+^2 v_{-1}(t) = 0,$$

which is equivalent to

$$v_{-1}(t) = 2v_0(t) - v_{+1}(t).$$

We eliminate the ghost point and thus calculate on the boundary

$$v_{0t}(t) = \frac{v_{+1}(t) - v_{-1}(t)}{2h} = \frac{v_1(t) - v_0(t)}{h} = D_+ v_0(t).$$

We can define a fourth order approximation by

$$v_{\nu t}(t) = \frac{4}{3} D_0(h) \, v_\nu(t) - \frac{1}{3} D_0(2h) \, v_\nu(t),$$

where

$$D_0(h) \, v_\nu(t) = \frac{v_{\nu+1}(t) - v_{\nu-1}(t)}{2h}$$

and

$$D_0(2h) \, v_\nu(t) = \frac{v_{\nu+2}(t) - v_{\nu-2}(t)}{4h}.$$

Here we need two ghost points x_{-1} and x_{-2} and thus we need further boundary conditions.

We consider two techniques to investigate for what p the approximation (3.2.4) with boundary conditions (3.2.5) is stable.

3.2.1 Energy Estimates

We use the discrete scalar product and norm

$$(u, v)_h = h \sum_{\nu=0}^{\infty} \overline{u}_\nu v_\nu, \quad \|u\|_h^2 = (u, u)_h.$$

Corresponding to (3.2.2) we have

Lemma 3.2.1 *For any gridfunctions u, v we have*

$$(u, D_0 v)_h = -(D_0 u, v)_h - \frac{1}{2} \left(\overline{u}_0 v_{-1} + \overline{u}_{-1} v_0 \right).$$

Proof. By definition

$$\begin{aligned}(u, D_0 v)_h &= \frac{1}{2}\sum_{\nu=0}^{\infty} \bar{u}_\nu v_{\nu+1} - \frac{1}{2}\sum_{\nu=0}^{\infty} \bar{u}_\nu v_{\nu-1} \\ &= \frac{1}{2}\sum_{\nu=1}^{\infty} \bar{u}_{\nu-1} v_\nu - \frac{1}{2}\sum_{\nu=-1}^{\infty} \bar{u}_{\nu+1} v_\nu \\ &= \frac{1}{2}\sum_{\nu=0}^{\infty} \left(\bar{u}_{\nu+1} - \bar{u}_{\nu-1}\right) v_\nu - \frac{1}{2}\left(\bar{u}_0 v_{-1} + \bar{u}_{-1} v_0\right) \\ &= -\left(D_0 u, v\right)_h - \frac{1}{2}\left(\bar{u}_0 v_{-1} + \bar{u}_{-1} v_0\right). \qquad \square\end{aligned}$$

If we consider

$$\frac{\partial}{\partial t}\|v\|_h^2 = (v, v_t)_h + (v_t, v)_h = (v, D_0 v)_h + (D_0 v, v)_h,$$

we obtain by Lemma 3.2.1 that

$$\frac{\partial}{\partial t}\|v\|_h^2 = -\frac{1}{2}\left(\bar{v}_0 v_{-1} + \bar{v}_{-1} v_0\right). \tag{3.2.6}$$

Since the right-hand side of Equation (3.2.6) is not a square, it is in general not negative. For the special boundary condition

$$v_{-1} = v_0, \quad \text{we have} \quad -\frac{1}{2}\left(\bar{v}_0 v_{-1} + \bar{v}_{-1} v_0\right) = -|v_0|^2,$$

and the approximation is stable. For $p = 2$ we change the scalar product $(u, v)_h$ to

$$\widetilde{(u, v)}_h = \frac{1}{2}\bar{u}_0 v_0 + h\sum_{\nu=1}^{\infty} \bar{u}_\nu v_\nu.$$

In the same way as Lemma 3.2.1 one can derive the identity

$$\widetilde{(D_0 u, v)}_h = -\widetilde{(u, D_0 v)}_h - \frac{1}{2}\left(\bar{u}_0 v_0 + u_0 \bar{v}_0\right).$$

for all gridfunctions u, v which satisfy (3.2.4) with $p = 2$. Thus for $p = 2$

$$\frac{\partial}{\partial t}\|v\|_h^2 = -|v_0|^2$$

and the approximation is stable.

Unfortunately we cannot use the same trick for $p > 2$. Without proof, we state the following theorem.

Theorem 3.2.1 *For $p > 2$, there is no way to change the norm such that we get an energy estimate.*

3.2.2 Estimates by using Laplace Transform

We have the same test as for the continuous case. Denote by $\|\cdot\|_h$ the norm of a gridfunction, defined by

$$\|\varphi\|_h^2 = h \sum_{\nu=0}^{\infty} |\varphi(x_\nu)|^2.$$

Lemma 3.2.2 *The approximation (3.2.4) is not stable, if it has for some $s = s_0$, $h = h_0$ a nontrivial solution of the type*

$$v(x_\nu, t) = e^{st} \varphi(x_\nu), \tag{3.2.7}$$

with $\operatorname{Re} s > 0$ *satisfying the boundary condition*

$$(hD_+)^p \varphi(x_{-1}) = 0. \tag{3.2.8}$$

Proof. If (3.2.7) is a nontrivial solution of our difference approximation for some $s = s_0$, $\operatorname{Re} s_0 > 0$, $h = h_0$, then it satisfies

$$s_0 \varphi(x_\nu) = D_0 \varphi(x_\nu)$$

or equivalently

$$2h_0 s_0 \varphi(x_\nu) = \varphi(x_{\nu+1}) - \varphi(x_{\nu-1}). \tag{3.2.9}$$

If we replace s_0, h_0 by αs_0, h_0/α we still have a solution which grows like $e^{\operatorname{Re}\,\alpha s_0 t}$. Thus we can construct for $h \to 0$ solutions which grow arbitrary fast. □

To solve the difference equation (3.2.9) we choose the ansatz

$$\varphi(x_\nu) = \kappa^\nu,$$

which leads to the characteristic equation

$$2hs = \kappa - \frac{1}{\kappa}. \tag{3.2.10}$$

Lemma 3.2.3 *For* $\operatorname{Re} s > 0$, *there is no solution κ of the characteristic equation (3.2.10) with $|\kappa| = 1$. There is exactly one solution κ_1 with $|\kappa_1| < 1$ and one solution κ_2 with $|\kappa_2| > 1$. Also, there is a constant $\delta > 0$ such that $|\kappa_1 - 1| \geq \delta$ for* $\operatorname{Re} s \geq 0$.

Proof. Assume that $\operatorname{Re} s > 0$ and $|\kappa| = 1$, i.e. $\kappa = e^{i\xi}$ for some $\xi \in \mathbb{R}$. Then we introduce $\kappa = e^{i\xi}$ into equation (3.2.10) and obtain

$$2hs = e^{i\xi} - e^{-i\xi} = 2i \sin \xi,$$

in contradiction to the assumption that $\operatorname{Re} s > 0$. For $|2hs| \gg 1$, Equation 3.2.10 tells us that $|\kappa_1| \sim |2hs|^{-1} \ll 1$, $|\kappa_2| \sim |2hs| \gg 1$. Since κ_1, κ_2 are continuous functions of hs, we have for all $\operatorname{Re} s > 0$ that $|\kappa_1| < 1$, $|\kappa_2| > 1$. The only possibility that $\kappa_1 \to 1$ occurs when $s \to 0$. We expand the solution κ around small hs. The quadratic equation

$$\kappa^2 - 2hs\kappa - 1 = 0$$

is solved by

$$\kappa = hs \pm \sqrt{h^2 s^2 + 1}.$$

Thus

$$\kappa_1 \sim -1 + hs + \ldots$$

The proof is complete. □

Thus φ_ν is of the form

$$\varphi_\nu = \sigma_1 \kappa_1^\nu + \sigma_2 \kappa_2^\nu.$$

Since we required that $\|\varphi\|_\nu < \infty$, it follows that $\sigma_2 = 0$. Thus

$$\varphi_\nu = \sigma_1 \kappa_1^\nu. \tag{3.2.11}$$

If we plug (3.2.11) into Equation (3.2.8), we obtain

$$(hD_+)^p \varphi_{-1} = \sigma_1 \frac{(\kappa_1 - 1)^p}{\kappa_1} = 0,$$

which can only be satisfied, if $\sigma_1 = 0$. In this example the test is negative, but it is not guaranteed that the approximation is stable.

Consider the approximation

$$\begin{aligned} v_{\nu t}(t) &= D_0 v_\nu(t) + F_\nu(t), \\ (hD_+)^p v_{-1}(t) &= g(t), \\ v_\nu(0) &= 0. \end{aligned} \tag{3.2.12}$$

Definition 3.2.1 *Consider the approximation 3.2.12 with $g \equiv 0$. We call the problem stable in the generalized sense, if for all $h < h_0$ there is a unique solution that satisfies the estimate*

$$\int_0^\infty e^{-2\eta t} \|v\|_h^2 dt \le K(\eta) \int_0^\infty e^{-2\eta t} \|F\|_h^2 dt, \tag{3.2.13}$$

for all $\eta > \eta_0$. Here η_0 and $K(\eta)$ are constants that do not depend on F and, furthermore,

$$\lim_{\eta \to \infty} K(\eta). \tag{3.2.14}$$

To obtain a test which guarantees stability in the generalized sense we again introduce correspondingly to the continuous case the concept of boundary stability.

Definition 3.2.2 *The approximation (3.2.12) is boundary stable, if we can estimate the solution $v_\nu(t)$ for all points which are involved in the boundary condition*

$$|\hat{v}_\nu(s)| \leq K|\hat{g}(s)|, \quad \nu = -1, \dots, p-1. \tag{3.2.15}$$

Lemma 3.2.4 *The approximation*

$$\begin{aligned} w_{\nu t}(t) &= D_0 w_\nu(t), \\ (hD_+)^p w_{-1}(t) &= g(t), \ \|w\|_h < \infty \\ w_\nu(0) &= 0, \end{aligned} \tag{3.2.16}$$

is boundary stable.

Proof. We Laplace transform Equations (3.2.16) and obtain

$$\begin{aligned} s\hat{w}(s) &= D_0 \hat{w}_\nu(s), \\ (hD_+)^p \hat{w}_{-1}(s) &= \hat{g}(s), \ \|\hat{w}\|_h < \infty \\ \hat{w}_\nu(0) &= 0. \end{aligned} \tag{3.2.17}$$

The general solution of (3.2.17) is

$$\hat{w}_\nu(s) = \sigma(s)\kappa_1^\nu.$$

$\sigma(s)$ is determined by the boundary condition i.e.

$$\frac{\sigma(s)(\kappa_1 - 1)^p}{\kappa_1} = \hat{g}(s).$$

Since $|\kappa_1 - 1| \geq \delta$, it follows that $\sigma(s)$ is bounded by $\hat{g}(s)$

$$|\sigma(s)| \leq K|\hat{g}(s)|.$$

Thus we have bounds for all points which are involved in the boundary condition

$$|\hat{w}_\nu(s)| \leq K|\hat{g}_\nu(s)|, \quad \nu = -1, \dots, p-1,$$

and Lemma 3.2.4 is proved. □

We shall now use Lemma 3.2.4 to prove the following theorem:

Theorem 3.2.2 *The approximation (3.2.12) is stable in the generalized sense.*

Proof. Similar to the continuous case we need to derive the estimate

$$\|\hat{v}\|_h \le K(\eta)\,\|\hat{F}\|_h, \quad K(\eta) \le \frac{K_0}{\eta}.$$

We first solve an auxiliary problem with boundary conditions for which we obtain an energy estimate.

$$\begin{aligned} s\hat{w}_\nu(s) &= D_0\hat{w}_\nu(s) + \hat{F}_\nu(s), \\ \hat{w}_{-1}(s) &= \hat{w}_0(s). \end{aligned} \tag{3.2.18}$$

By Lemma 3.2.1

$$\begin{aligned} \operatorname{Re} s\|\hat{w}\|_h^2 &= \operatorname{Re}(\hat{w}, D_0\hat{w})_h + \left(\hat{w}, \hat{F}\right)_h \\ &\le -\tfrac{1}{2}\left(|\hat{w}_0|^2 + |\hat{w}_{-1}|^2\right) + \operatorname{Re}\|\hat{w}\|_h\|\hat{F}\|_h. \end{aligned}$$

Therefore

$$2\ \operatorname{Re} s\|\hat{w}\| \le \|\hat{F}\|_h$$

and

$$\frac{1}{2}\left(|\hat{w}_0|^2 + |\hat{w}_{-1}|^2\right) \le \frac{1}{\operatorname{Re} s}\|\hat{F}\|_h^2.$$

Using the difference equation

$$\hat{w}_{\nu+1} = 2hs\hat{w}_\nu + \hat{w}_{\nu-1} + 2h\hat{F}_\nu,$$

we can obtain estimates of the same type for any ν.

$$|\hat{w}_\nu|^2 \le \frac{K_\nu}{\operatorname{Re} s}\|\hat{F}\|_h^2. \tag{3.2.19}$$

Next we consider the difference $\hat{y}_\nu(s) = \hat{v}_\nu(s) - \hat{w}_\nu(s)$ of the solution of Equation (3.2.12) and the solution of (3.2.18). We observe that $\hat{y}_\nu(s)$ satisfies

$$\begin{aligned} s\hat{y}_\nu(s) &= D_0\,\hat{y}_\nu(s), \\ (hD_+)^p\,\hat{y}_{-1}(s) &= (hD_+)^p\,\hat{w}_{-1}(s). \end{aligned}$$

It follows from Lemma 3.2.4 that

$$|\hat{y}_\nu(t)| \le K|(hD_+)^p\hat{w}_{-1}(s)|, \ \nu = -1, \dots, p-1. \tag{3.2.20}$$

From Equation (3.2.19) it follows that

$$|(hD_+)^p\hat{w}_{-1}(s)| \le \frac{K}{\operatorname{Re} s}\|\hat{F}\|_h^2. \tag{3.2.21}$$

Equations (3.2.20) and (3.2.21) together give an estimate $\hat{y}_\nu(s)$ near the boundary. To get an estimate for $\hat{y}_\nu(s)$ in the interior of the domain, we use summation by parts. By Lemma 3.2.1

$$\operatorname{Re} s\|\hat{y}_\nu\|_h^2 = \operatorname{Re}(\hat{y}_\nu, D_0\hat{y}_\nu) = -\frac{1}{2}\left(\bar{\hat{y}}_0\hat{y}_{-1} + \bar{\hat{y}}_{-1}\hat{y}_0\right) \leq \hat{K}|\hat{g}|^2.$$

Thus we have

$$\|\hat{y}_\nu\|_h^2 \leq \frac{K}{\operatorname{Re} s^2}\|\hat{F}\|_h^2$$

and the proof is complete. □

Next we formulate a general version of Theorem 3.2.2 for the one-dimensional case. The generalization to multiple space dimensions goes via Fourier transform in the tangential direction. Consider the one-dimensional system

$$u_t(x,t) = Au_x(x,t), \tag{3.2.22}$$

where we assume that the matrix A is symmetric, $A = A^*$. We approximate (3.2.22) by

$$v_{\nu t}(t) = \sum_{j=-r}^{s} C_j E^j v_\nu(t) + F_\nu(t). \tag{3.2.23}$$

The difference operator has to be *consistent*, i.e.

$$\sum C_j = 0 \quad \text{and} \quad \sum jC_j = A.$$

We need r conditions at the boundary

$$v_\mu(t) = \sum_{j=0}^{p} D_{\mu j}v_j(t) + g_\mu(t), \quad \mu = -1, -2, \ldots, -r.$$

We assume that the pure Cauchy problem is stable. Then it follows, applying our test, that the number of boundary conditions is equal to the number of κ_j with $|\kappa_j| < 1$, where κ_j denote the zeros of the characteristic equation.

Theorem 3.2.3 *If the problem is boundary stable, then it is stable in the generalized sense and we have the following estimate*

$$\int_0^\infty e^{-2\eta t}\|v\|_h dt \leq \int_0^\infty e^{-2\eta t}\left(\|F\|_h^2 + |g|^2\right) dt. \tag{3.2.24}$$

3.2.3 Error Estimates

The above estimates are very powerful to find error estimates. Consider the differential equation

$$\begin{aligned} u_t(x,t) &= P(\partial/\partial x)\, u(x,t) + F(x,t), \\ u(x,0) &= f(x), \\ \tilde{B}u(0,t) &= h(t). \end{aligned} \tag{3.2.25}$$

we approximate it by

$$\begin{aligned} v_{\nu t}(t) &= Q(E)\, v_\nu(t) + F_\nu(t), \\ v_\nu(x,0) &= f_\nu, \\ Bv_\nu(t) &= g(t). \end{aligned} \tag{3.2.26}$$

We assume that Equation (3.2.25) has a smooth solution and introduce it into Equation (3.2.26) to obtain, by truncation error analysis,

$$\begin{aligned} u_{\nu t}(t) &= Q(E)\, u_\nu(t) + F_\nu(t) + h^p R_\nu(t), \\ u_\nu(0) &= f_\nu, \\ Bu_0(t) &= g(t) + h^p r(t). \end{aligned}$$

The difference $e = u - v$ satisfies the same difference equations with F, f, g replaced by zero. The error estimate follows from (3.2.24).

3.3 Method of Lines

If we discretize in space and leave time continuous, we obtain a system of ordinary differential equations. We can solve this system by a standard method, for example by a Runge-Kutta method. This technique is then called *method of lines.*

We assume that the approximation

$$v_t(x,t) = \frac{1}{h} Qv(x,t) + F(x,t) \tag{3.3.1}$$

is stable in the generalized sense. Q denotes a difference operator, which already contains the boundary conditions. To solve the resulting system of ordinary differential equations, we use a Runge-Kutta method.

$$w(t+k) = P\left(\frac{k}{h}Q\right) w(t) + kG, \tag{3.3.2}$$

where P is a polynomial in $\frac{k}{h}Q$, G is a polynomial in F and $\frac{k}{h}$ is bounded. We recall a definition from the standard ODE-theory.

Definition 3.3.1 *If we apply the Runge-Kutta method (3.3.2) to the problem*

$$y'(t) = \lambda y(t),$$

we obtain the difference equation

$$w(t+k) = P(kA)\, w(t).$$

Let $\mu = \lambda k$. The stability region Ω of the method (3.3.2) is then defined by

$$\mu \in \Omega \Leftrightarrow |P(\mu)| < 1.$$

Definition 3.3.2 *We call a method (3.3.2) uniformly stable, if the stability region Ω contains a half-circle $|\mu| \leq R_1$ and $\operatorname{Re}\mu < 0$.*

In [3] the following theorem is proved

Theorem 3.3.1 *If the underlying ODE-method is uniformly stable, then the totally disretized PDE-method is stable in the generalized sense, if*

$$\left|\frac{k}{h}Q\right| < R_1 \tag{3.3.3}$$

Convenient ODE-solvers to construct a method of lines are for example Runge-Kutta of order 3 and 4, Adams-Moulton of order 1, 2, 5 and 6, Adams-Bashforth of order 3, 4, 7 and 8 and predictor-corrector methods.

The leapfrog method is not uniformly stable and there can be stability problems if one is not careful with the approximation of the boundary conditions.

3.4 Remarks

For the material in Section 3.1, consult Gustafsson, Kreiss and Oliger, Chapter 2, for the material in Section 3.2, consult Gustafsson, Kreiss and Oliger, Chapter 11.

Chapter 4

Nonlinear Problems

4.1 General Discussion

In this section we shall discuss nonlinear problems. We are interested in smooth solutions. There is no general theory for nonlinear differential equations available. Instead, we ask the following questions. Assume that we know a solution U for a particular set of data. Is the problem still solvable if we make small perturbations of the data? Does the solution depend continuously on the perturbation, i.e., do small perturbations in the data generate small changes in the solution?

We can linearize the nonlinear equations around the known solution U and we will see that properties of this linear system often determine the answer to the above questions.

In practice, one often solves nonlinear problems numerically without having any knowledge whether the differential equations have a solution. If the numerical solution is smooth in the sense that it varies slowly with respect to the mesh, then we can interpolate the numerical solution. The interpolant solves a nearby problem and the solution of the original problem can be considered as a perturbation of the numerically constructed solution. Therefore, the above questions are of interest.

4.2 Initial Value Problems for Ordinary Differential Equations

We start with a simple model problem

$$\begin{aligned} y_t &= \alpha y + \varepsilon y^2, \quad t \geq 0, \\ y(0) &= y_0. \end{aligned} \tag{4.2.1}$$

Here α, ε are real constants with $0 < \varepsilon \ll 1$. We assume also that $y_0 > 0$ and we shall calculate the solution explicitly. Introducing a new variable by

$$y = e^{\alpha t}\tilde{y}$$

gives us

$$\begin{aligned} \tilde{y}_t &= \varepsilon e^{\alpha t}\tilde{y}^2, \\ \tilde{y}(0) &= y_0. \end{aligned}$$

Therefore,

$$\frac{1}{y_0} - \frac{1}{\tilde{y}(t)} = \int_0^t \frac{\tilde{y}_\xi}{\tilde{y}^2} d\xi = \varepsilon \int_0^t e^{\alpha\xi} d\xi = \varepsilon\psi(t,\alpha),$$

where

$$\psi(t,\alpha) = \begin{cases} \frac{e^{\alpha t}-1}{\alpha} & \text{if } \alpha \neq 0, \\ t & \text{if } \alpha = 0. \end{cases}$$

Solving for $\tilde{y}(t)$ gives us

$$\tilde{y}(t) = \frac{1}{\frac{1}{y_0} - \varepsilon\psi(t,\alpha)}. \tag{4.2.2}$$

There are three different regimes.

If $\alpha > 0$, then the solution blows up and the blow up time

$$T = \frac{1}{\alpha}\log\left(1 + \frac{\alpha}{y_0\varepsilon}\right).$$

Thus, it does not help very much to decrease ε. T increases only logarithmically.

If $\alpha = 0$, then $T = \frac{1}{y_0\varepsilon}$ and the blow up time is linear in $1/\varepsilon$.

If $\alpha < 0$, there is no blow up for sufficiently small ε.

The above discussion shows that the sign of α is the dominating factor determining the behavior of the solution. If $\alpha < 0$, we can, for sufficiently small ε, neglect the nonlinear term. This is also true if $\alpha = 0$, provided the time interval we consider is not too large.

The same type of results holds for general equations

$$\begin{aligned} y_t &= \alpha y + \varepsilon F(y,t), \\ y(0) &= y_0, \end{aligned} \tag{4.2.3}$$

where $F(y,t)$ denotes the nonlinear term.

We shall now solve a nonlinear equation

$$\begin{aligned} y_t &= f(y,t), \\ y(0) &= y_0, \end{aligned} \tag{4.2.4}$$

using the forward Euler method. Let $k > 0$ denote the gridsize and $v_n = v(nk)$ the approximation of y on the grid. Euler's forward method can be written as

$$\begin{aligned} v_{n+1} &= v_n + kf(v_n, t_n), \\ v_0 &= y_0. \end{aligned} \tag{4.2.5}$$

We calculate v_n in some time interval $0 \leq t \leq T$ and want to decide whether the numerical solution v has anything to do with the analytic solution y.

It is well known that we can interpolate the discrete gridfunction v by splines such that the resulting interpolant $\varphi = \operatorname{Int} v$ belongs to $C^p(0,T)$ and

$$\sum_{j=0}^{p} \left| \frac{d^j \varphi(\cdot)}{dt^j} \right|_\infty \leq K_p \sum_{j=0}^{p} |D_k^j v|_{k,\infty}.$$

Here $D_k^j v_\nu$ denote the divided differences of order j and

$$\left| \frac{d^j \varphi(\cdot)}{dt^j} \right|_\infty = \max_t \left| \frac{d^j \varphi(\cdot)}{dt^j} \right|, \quad |D_k^j v|_{k,\infty} = \max_\nu |D_k^j v_\nu|.$$

We can choose p arbitrarily. The constants K_p increase with p and are of order 2^p (see [1]).

We want to show that $\varphi(t)$ solves a nearby differential equation. We need

Lemma 4.2.1 *Consider a time interval* $0 \leq t \leq T$ *which we cover by a grid* $t_\nu = \nu k,\ \nu = 0, 1, 2, \ldots, N;\ Nk = T$. *Let* $F \in C^1$ *be a function with*

$$\max_{0 \leq \nu \leq N} |F(t_\nu)| = \delta.$$

Then

$$|F(\cdot)|_\infty \leq \delta + k|F_t(\cdot)|_\infty, \quad |F(\cdot)|_\infty = \max_{0 \leq t \leq T} |F(t)|.$$

Proof. For every $t \in (0, T)$, there are gridpoints $t_\nu, t_{\nu+1}$ such that

$$t_\nu \leq t < t_{\nu+1}.$$

Therefore,

$$F(t) = F(t_\nu) + \int_{t_\nu}^{t} F_t(\xi)d\xi,$$

and the lemma follows. □

Without proof we state the following generalization.

Lemma 4.2.2 *If* F *in Lemma 4.2.1 belongs to* C^p, *then there is a constant* C_p *such that*

$$|F(t)|_\infty \leq \delta + C_p k^p \left| \frac{d^p F(\cdot)}{dt^p} \right|_\infty.$$

We assume now that the interpolant φ belongs to C^2. Then $\varphi_t - f(\varphi, t) \in C^1$ and we have bounds for $\big(\varphi_t - f(\varphi, t)\big)_t = \varphi_{tt} - f_\varphi \varphi_t - f_t$.

For every gridpoint

$$
\begin{aligned}
&|\varphi_t(t_\nu) - f\big(\varphi(t_\nu), t_\nu\big)| \\
&\quad = |\varphi_t(t_\nu) - f\big(\varphi(t_\nu), t_\nu\big) - \frac{\varphi(t_{\nu+1}) - \varphi(t_\nu)}{k} + f\big(\varphi(t_\nu), t_\nu\big)| \\
&\quad = |\varphi_t(t_\nu) - \frac{\varphi(t_{\nu+1}) - \varphi(t_\nu)}{k}| \le \frac{k}{2}|\varphi_{tt}(\cdot)|_\infty.
\end{aligned}
$$

Therefore, by Lemma 4.2.1, we have, for arbitrary t,

$$\varphi_t = f\big(\varphi(t), t\big) + kR(t), \tag{4.2.6}$$

where

$$|R(t)| \le \frac{1}{2}|\varphi_{tt}(\cdot)|_\infty + |\big(\varphi_t - f(\varphi, t)\big)_t|_\infty.$$

$R(t)$ is, essentially, the truncation error evaluated at the interpolant. If $\varphi \in C^p$, then $R(t) \in C^{p-2}$ and we have bounds for the derivatives of R in terms of the divided differences of the numerical solution. Thus, if the numerical solution looks "smooth" which means that the divided differences are bounded, then the interpolant solves a nearby differential equation.

This is as close as numerical methods can get us to the true solution. If we want to know how close we are to the true solution $y(t)$, we have to use perturbation theory. We make the change of variables $y(t) = \varphi(t) + k\tilde{y}(t)$ and (4.2.4) becomes

$$
\begin{aligned}
\tilde{y}_t(t) &= \tfrac{f(\varphi + k\tilde{y}, t) - f(\varphi, t)}{k} - R(t) \\
&= \tfrac{\partial f(\varphi, t)}{\partial \varphi}\tilde{y} + kg(\tilde{y}, t) - R(t), \\
\tilde{y}(0) &= 0.
\end{aligned} \tag{4.2.7}
$$

Here $g(\tilde{y}, t)$ is quadratic in $\tilde{y}$. As in the model problem, the behavior of the linearized problem

$$
\begin{aligned}
\tilde{\tilde{y}}_t &= \tfrac{\partial f(\varphi, t)}{\partial \varphi}\tilde{\tilde{y}} - R(t), \\
\tilde{\tilde{y}}(0) &= 0,
\end{aligned} \tag{4.2.8}
$$

tells us how long $\tilde{y}(t)$ stays bounded. If the solution operator of (4.2.8) decays exponentially, then, for sufficiently small k, $\tilde{y}(t)$ stays bounded for all times. On the other hand, if the solution operator grows exponentially, the blow up can occur at $t = \mathcal{O}(\log(\frac{1}{k})$.

For complicated problems, one has no analytic knowledge of the behavior of the solution operator. Therefore, one relies on numerical perturbation calculations.

Instead of the Euler method, we could have used a higher order method like the fourth order Runge-Kutta method. In that case the forcing in the equation (4.2.6) would be of order k^4.

The procedure can also be used for partial differential equations but the estimates of the derivatives of the interpolant become rather complicated.

4.3 Existence Theorems for Nonlinear Partial Differential Equations

Consider the Cauchy problem for a quasilinear first order system

$$\begin{aligned} \frac{\partial \tilde{u}}{\partial t} &= P(x,t,\tilde{u},\tfrac{\partial}{\partial x})\tilde{u} + F_1(x,t), \\ \tilde{u}(x,0) &= f_1(x). \end{aligned} \tag{4.3.9}$$

Even if we assume that all coefficients and data are smooth functions of all variables, there is no global existence theory available. The only general results are of local character which can be phrased in the following way. Assume that we know that a nearby problem

$$\begin{aligned} \frac{\partial U}{\partial t} &= P(x,t,U,\tfrac{\partial}{\partial x})U + F_1(x,t) - \varepsilon F(x,t), \\ U(x,0) &= f_1(x) - \varepsilon f(x), \end{aligned} \tag{4.3.10}$$

has a smooth solution in some time interval $0 \le t \le T$. Here $\varepsilon > 0$ is a small constant. Can we infer that, for sufficiently small ε, also the original problem (4.3.9) has a solution in the same time interval and that $|U - u| = \mathcal{O}(\varepsilon)$? Using the arguments of the previous section, U can often, at least in principle, be obtained by interpolating a numerical solution of the problem.

The change of variables

$$\tilde{u} = U + \varepsilon u$$

leads to the system

$$\begin{aligned} \frac{\partial u}{\partial t} &= P_0(x,t,\tfrac{\partial}{\partial x})u + \varepsilon P_1(x,t,u,\tfrac{\partial}{\partial x})u + F(x,t), \\ u(x,0) &= f(x). \end{aligned} \tag{4.3.11}$$

Here P_0 denotes the linear operator which one obtains by linearizing $P(x,t,\tilde{u},\partial/\partial x)\tilde{u}$ around U. A natural assumption is that the linear problem

$$\begin{aligned} \frac{\partial w}{\partial t} &= P_0(x,t,\tfrac{\partial}{\partial x})w + F(x,t), \\ w(x,0) &= f(x), \end{aligned} \tag{4.3.12}$$

is a well-posed problem. We shall now give arguments that, under reasonable conditions, this assumption guarantees that also (4.3.11) has a solution, provided ε is sufficiently small. We start with an example and consider

$$\begin{aligned} u_t &= \alpha u + \varepsilon u u_x + F(x,t), \\ u(x,0) &= f(x). \end{aligned} \tag{4.3.13}$$

Here α, ε are real constants and $f, F \in C^\infty$ are real smooth functions which are 2π-periodic in x. We are interested in real solutions which are also 2π-periodic in x.

The most important tool to derive existence theorems are a priori estimates, i.e., we assume that there is a smooth solution and we derive estimates of u and its derivatives in terms of f and F and their derivatives. Once one has obtained these estimates, existence follows. Here we shall only derive the a priori estimates. We refer to the literature how to use them for existence theorems.

As before, (u,v), $\|u\|^2 = (u,u)$ denote the L_2-scalar product and norm, here with respect to the interval $0 \le x \le 2\pi$. Multiplying (4.3.13) with u and integrating gives

$$\frac{1}{2}(u,u)_t = \alpha\|u\|^2 + (u, uu_x) + (u, F). \tag{4.3.14}$$

Assuming a periodic problem, integration by parts implies

$$(u, uu_x) = (u^2, u_x) = -2(uu_x, u) = -2(u, uu_x),$$

i.e.,

$$(u, uu_x) = 0.$$

Therefore, (4.3.14) gives us

$$\frac{1}{2}\|u\|_t^2 \le \alpha\|u\|^2 + \|u\|\,\|F\|,$$

i.e., using $\|u\|_t^2 = 2\|u\| \cdot \|u_t\|$,

$$\|u\|_t \le \alpha\|u\| + \|F\|, \quad \|u(\cdot,0)\| = \|f\|.$$

Therefore,

$$\begin{aligned} \|u(\cdot,t)\| &\le e^{\alpha t}\|f\| + \textstyle\int_0^t e^{\alpha(t-\xi)}\|F(\cdot,\xi)\|\,d\xi \\ &\le e^{\alpha t}\|f\| + \max_{0\le\xi\le t}\|F(\cdot,\xi)\|\,\psi(\alpha,t), \end{aligned} \tag{4.3.15}$$

where again

$$\psi(\alpha,t) = \begin{cases} \dfrac{e^{\alpha t}-1}{\alpha} & \text{for } \alpha \ne 0, \\ t & \text{for } \alpha = 0 \end{cases}.$$

As in the ODE case, the sign of α determines whether the solution grows or stays bounded.

To obtain bounds for $v = u_x$ we differentiate (4.3.13) with respect to x and obtain

$$\begin{aligned} v_t &= \alpha v + \varepsilon u v_x + \varepsilon v^2 + F_x, \\ v(x,0) &= f_x. \end{aligned} \tag{4.3.16}$$

Therefore,

$$\frac{1}{2}\|v\|_t^2 = \alpha\|v\|^2 + \varepsilon(v, uv_x) + \varepsilon(v, v^2) + (v, F_x).$$

Since

$$(v, uv_x) = -(v_x u, v) - (v, v^2), \quad (v, v^2) \le |v|_\infty \|v\|^2,$$

it follows that

$$(v, uv_x) = -\frac{1}{2}(v, v^2)$$

and

$$\frac{1}{2}\|v\|_t^2 \le \alpha\|v\|^2 + \frac{|\varepsilon|}{2}|v|_\infty\|v\|^2 + \|v\|\,\|F_x\|,$$

i.e.,

$$\begin{aligned} \|v\|_t &\le \alpha\|v\| + \tfrac{\varepsilon}{2}|v|_\infty\|v\| + \|F_x\|, \\ \|v(\cdot,0)\| &= \|f_x\|. \end{aligned} \tag{4.3.17}$$

Since $|v|_\infty$ cannot be estimated in terms of $\|v\|$, we cannot use (4.3.17) directly to estimate v.

Differentiating (4.3.13) gives us an equation for $w = u_{xx}$

$$w_t = \alpha w + \varepsilon u w_x + 3\varepsilon v w + F_{xx}.$$

Therefore,

$$\frac{1}{2}\|w\|_t^2 = \alpha\|w\|^2 + \varepsilon(w, uw_x) + 3\varepsilon(w, vw) + (w, F_{xx}).$$

Since

$$(w, u\,w_x) = -(w_x, u\,w) - (w, v\,w),$$

we obtain

$$\begin{aligned} \|w\|_t &\le \alpha\|w\| + 4|\varepsilon|\,|v|_\infty\|w\| + \|F_{xx}\| \\ &\le \alpha\|w\| + 2|\varepsilon|\,|v|_\infty^2 + 2|\varepsilon|\,\|w\|^2 + \|F_{xx}\|. \end{aligned} \tag{4.3.18}$$

We now derive a Sobolev inequality to estimate $|v|_\infty$ in terms of $\|v\|, \|w\|$. Let x_1, x_0 be two points with

$$|v|_\infty = |v(x_1)|, \quad \min_{0\le x\le 2\pi} = |v(x)| = |v(x_0)|,$$

then

$$\int_{x_0}^{x_1} v\,w\,dx = \int_{x_0}^{x_1} u_x u_{xxx} dx = u_x^2(x_1) - u_x^2(0) - \int_{x_0}^{x_1} u_x u_{xx} dx.$$

Since

$$\min_{0\le x\le 2\pi} |v(x)|^2 \le \frac{1}{2\pi}\|v\|^2,$$

we obtain

$$|v|_\infty^2 \le \frac{1}{2\pi}\|v\|^2 + \frac{1}{2}\|v\|\,\|w\| \tag{4.3.19}$$

and (4.3.18) becomes

$$\begin{aligned} \|w\|_t &\le \alpha\|w\| + |\varepsilon|(\tfrac{1}{\pi}\|v\|^2 + \|v\|\,\|w\| + 2\|w\|^2) + \|F_{xx}\|, \\ \|w(\cdot,0)\| &= \|f_{xx}\|. \end{aligned} \tag{4.3.20}$$

(4.3.17) and (4.3.20) represent a closed system of differential inequalities for $\|v\|$ and $\|w\|$. If we replace the inequality sign by the equality sign, we obtain a system of differential equations which majorizes the inequalities. This system is of the same form as the model problem in the previous section. The blow up time, if any, depends on the sign of α and the size of $|\varepsilon|$.

There are no difficulties to estimate higher derivatives. They exist as long as u_x, u_{xx} stay bounded.

The above estimates can be generalized to rather general mixed hyperbolic-parabolic systems. Consider, for example, the Cauchy problem for a quasilinear first order system

$$\begin{aligned} \tfrac{\partial u}{\partial t} &= P_0(x,t,\tfrac{\partial}{\partial x})u + \varepsilon P_1(x,t,u,\tfrac{\partial}{\partial x})u + F, \\ u(x,0) &= f. \end{aligned} \tag{4.3.21}$$

Here $F = F(x,t)$, $f = f(x,t)$, $u = u(x,t)$ are vector valued functions with n components depending on $x = (x_1,\dots,x_s) \in R_s$ and t.

$$P_0 = \sum_{j=1}^{s} A_j^{(0)}(x,t)\frac{\partial}{\partial x_j} + B^{(0)}(x,t), \quad P_1 = \sum_{j=1}^{s} A_j^{(1)}(x,t,u)\frac{\partial}{\partial x_j} + B^{(1)}(x,t,u)$$

are first order operators with symmetric matrix coefficients which depend smoothly on all variables.

Integration by parts shows that there is a constant α such that

$$(w, P_0 w) \le \alpha\|w\|^2 \quad \text{for all smooth } w.$$

We have

Theorem 4.3.1 *The system (4.3.21) has a smooth solution in some time interval $0 \le t \le T$. T depends on the sign of α and on f, F and ε.*

If $\alpha < 0$, then $T = \infty$ if ε is sufficiently small.
If $\alpha = 0$, then T is of the order $\mathcal{O}(\frac{1}{\varepsilon})$.
If $\alpha > 0$, then T is of the order $\mathcal{O}(\log(\frac{1}{\varepsilon}))$.

4.4 Perturbation Expansion

Consider the system (4.3.11) and assume that the linear problem (4.3.12) is a well posed problem, i.e., has a smooth solution $w(x,t)$. We expect that $w(x,t)$ is close to the solution of (4.3.11), at least in a sufficiently small time interval. Therefore, we make a change of variables

$$u = w + \varepsilon u_1$$

and obtain

$$\begin{aligned} \tfrac{\partial u_1}{\partial t} &= \left(P_0(x,t,\tfrac{\partial}{\partial x}) + \varepsilon P_{10}(x,t,\tfrac{\partial}{\partial x})\right) u_1 \\ &\quad + \varepsilon^2 P_2(x,t,u_1,\tfrac{\partial}{\partial x})u_1 + F_1(x,t), \\ &\quad u_1(x,0) = 0. \end{aligned}$$

Here P_{10} denotes the linear operator which we obtain by linearization of P_1. Again, we assume that the linear problem

$$\begin{aligned} \tfrac{\partial w_1}{\partial t} &= \left(P_0(x,t,\tfrac{\partial}{\partial x}) + \varepsilon P_{10}(x,t,\tfrac{\partial}{\partial x})\right) w_1 + F_1(x,t), \\ &\quad w_1(x,0) = 0, \end{aligned}$$

is a well posed problem. Thus, we can repeat the process and reduce the nonlinearity to any order in ε.

How useful this expansion is depends on the linear problems. If their solution operators decay exponentially, then, for sufficiently small ε, the expansion holds for all times. Also, we can estimate the rest term. Otherwise, the expansion might break down at some finite time T.

4.5 Convergence of Difference Methods

We shall solve the problem (4.3.13) by a difference method. For simplicity, we keep time continuous. We introduce a space grid $x_\nu = \nu h$, $h = 2\pi/N$, N a natural number, and denote by $v_\nu(t) = v(x_\nu, t)$ gridfunctions. We approximate (4.3.13) by the simplest centered approximation

$$\begin{aligned} \tfrac{dv_\nu}{dt} &= \alpha v_\nu + \varepsilon v_\nu D_0 v_\nu + F_\nu, \quad \nu = 0,1,2,\dots,N-1, \\ &\quad v_\nu(0) = f_\nu. \end{aligned} \qquad (4.5.22)$$

We consider a time interval where (4.3.13) has a smooth solution $u(x,t)$ and make the change of variables

$$v_\nu(t) = u_\nu(t) + h^2\tilde{v}_\nu(t). \tag{4.5.23}$$

(4.5.22) becomes

$$\begin{aligned} \frac{d\tilde{v}_\nu}{dt} &= (\alpha + \varepsilon D_0 u_\nu)\tilde{v}_\nu + \varepsilon u_\nu D_0\tilde{v}_\nu + \varepsilon h^2\tilde{v}_\nu D_0\tilde{v}_\nu + R_\nu(t), \\ \tilde{v}_\nu(0) &= 0, \end{aligned} \tag{4.5.24}$$

where

$$R_\nu = \frac{u_\nu D_0 u_\nu - u_\nu(u_x)_\nu}{h^2} = \frac{1}{6}(u_{xxx})_\nu + \mathcal{O}(h^2)$$

is the restriction of a smooth function to the net. Observe that the nonlinearity is of order $\mathcal{O}(\varepsilon h^2)$.

We shall now estimate the solution of (4.5.24). Corresponding to Section 3.2, we introduce a discrete scalar product and norm

$$(u,v)_h = \sum_{\nu=0}^{N} \bar{u}_\nu v_\nu h, \qquad \|u\|_h^2 = (u,u)_h.$$

(Here we assume that all functions are real, i.e., $\bar{u} = u$.)

Corresponding to Lemma 3.2.1, we have, for periodic functions,

$$(u, D_0 v)_h = -(D_0 v, u)_h. \tag{4.5.25}$$

Also, for any smooth function $a(x)$,

$$(u, aD_0 v)_h = (ua, D_0 v)_h = -\big(D_0(ua), v\big)_h.$$

Since

$$\begin{aligned} D_0 u_\nu a_\nu &= \frac{u_{\nu+1}a_{\nu+1} - u_{\nu-1}a_{\nu-1}}{2h} \\ &= a_\nu D_0 u_\nu + \tfrac{1}{2}\Big(u_{\nu+1}\frac{a_{\nu+1}-a_\nu}{h} - u_{\nu-1}\frac{a_\nu - a_{\nu-1}}{h}\Big), \end{aligned}$$

we obtain

$$(u, aD_0 v)_h = -(aD_0 u, v)_h + L \tag{4.5.26}$$

where

$$|L| \le |a_x|_\infty \|u\|_h \|v\|_h.$$

Now we multiply (4.5.24) by $\tilde{v}_\nu$ and obtain

$$\begin{aligned} \frac{1}{2}\frac{\partial}{\partial t}\|\tilde{v}\|_h^2 &\le (\alpha + \varepsilon|u_x|_\infty)\|\tilde{v}\|_h^2 + \varepsilon|(\tilde{v}, uD_0\tilde{v})_h| \\ &\quad + \varepsilon h^2|(\tilde{v}, \tilde{v}D_0\tilde{v})_h| + \|R\|_h^2. \end{aligned}$$

By (4.5.26),

$$(\tilde{v}, uD_0\tilde{v})_h = -(uD_0\tilde{v}, D_0\tilde{v})_h + L,$$

i.e.,

$$|(\tilde{v}, uD_0\tilde{v})_h| \le \frac{1}{2}|L| \le |u_x|_\infty \|\tilde{v}\|_h^2.$$

Since

$$|\tilde{v}_\nu|^2 \le \frac{1}{h}\|\tilde{v}\|_h^2, \quad \text{i.e.,} \quad |\tilde{v}|_\infty^2 \le \frac{1}{h}\|\tilde{v}\|_h^2,$$

it follows that

$$|(\tilde{v}, \tilde{v}D_0\tilde{v})_h| \le |D_0\tilde{v}|_\infty \|\tilde{v}\|_h^2 \le \frac{1}{h}|v|_\infty \|\tilde{v}\|_h^2 \le \frac{1}{h^{3/2}}\|\tilde{v}\|_h^3.$$

Therefore, we obtain

$$\frac{1}{2}\frac{\partial}{\partial t}\|\tilde{v}\|_h^2 \quad \le (\alpha + \varepsilon|u_x|_\infty)\|\tilde{v}\|_h^2 + \varepsilon h^{1/2}\|\tilde{v}\|_h^3 + \|R\|_h^2. \tag{4.5.27}$$

Thus, we again have the situation of Section 4.3. If $\alpha < 0$ and $\varepsilon h^{1/2}$ is sufficiently small, then the solution of (4.5.27) exists for all times and the error $\tilde{v} = v - u$ is of order $\mathcal{O}(h^2)$ for all times. If $\alpha > 0$, then we may have "blow up" at $T = \mathcal{O}(\log \frac{1}{\varepsilon h^{1/2}})$.

We can improve the situation slightly by using the perturbation expansion of Section 4.4. Neglecting the nonlinear term, we can consider (4.5.24) as a second order accurate approximation of

$$\begin{aligned} \frac{\partial w}{\partial t} &= (\alpha + \varepsilon u_x)w + \varepsilon u w_x + \tfrac{1}{6}u_{xxx}, \\ & w(x,0) = 0. \end{aligned}$$

Therefore, the substitution

$$\tilde{v}_\nu = w_\nu + h^2\tilde{\tilde{v}}_\nu$$

gives us

$$\begin{aligned} \frac{d\tilde{\tilde{v}}}{dt} &= (\alpha + \varepsilon D_0 u_\nu + \varepsilon h^2 D_0 w_\nu)\tilde{\tilde{v}}_\nu + \varepsilon(u_\nu + h^2 w_\nu)D_0\tilde{\tilde{v}}_\nu \\ &+\varepsilon h^4 \tilde{\tilde{v}}_\nu D_0 \tilde{\tilde{v}}_\nu + R_{1\nu}(t), \\ &\tilde{\tilde{v}}_\nu(0) = 0. \end{aligned}$$

Thus, we have reduced the nonlinearity to the order $\mathcal{O}(\varepsilon h^4)$. This process can be continued.

G. Strang [5] has used the procedure to prove convergence for very general problems.

4.6 Remarks

For the material in Section 4.3, consult Kreiss and Lorenz, Chapter 5, for the material in Section 4.5, consult Gustafsson, Kreiss and Oliger, Section 5.5.

Bibliography

[1] C. de Boor. How small can one make the derivatives of an interpolating function? *J. Approx. Theory*, 13:105–116, 1975.

[2] T. Hagström. Radiation boundary conditions for the numerical simulation of waves. *Acta Numerica*, pages 47–106, 1999.

[3] L. Wu H.O. Kreiss. On the stability definition of difference approximation for the boundary value problem. *Appl. Num. Math.*, 12:213–227, 1993.

[4] J.B. Keller M.J. Grote. Nonreflecting boundary conditions for time-dependent scattering. *J. Comp. Physics*, 127:52–65, 1996.

[5] G. Strang. Accurate partial difference methods ii. *Num. Math.*, 6:37–46, 1964.

Index